VESUVIUS

WONDERS OF THE WORLD

....

VESUVIUS

GILLIAN DARLEY

Harvard University Press
Cambridge, Massachusetts
2012

First published in Great Britain in 2011 by Profile Books Ltd,
3A Exmouth House, Pine Street, Exmouth Market,
London EC1R 0JH, U.K.

Typeset in Caslon by MacGuru Ltd

Designed by Peter Campbell

Library of Congress Cataloging-in-Publication Data

Darley, Gillian.
Vesuvius / Gillian Darley.
p. cm. — (Wonders of the world)
Originally published: London : Profile Books, 2011.
Includes bibliographical references and index.
ISBN 978-0-674-05285-7 (cloth : alk. paper)
1. Vesuvius (Italy)—History.
2. Volcanism—Italy—Vesuvius. I. Title.
QE523.V5D37 2012
945'.73—dc23
2011035047

To Susannah and Michael

'*Many men, many opinions, as one of the ancients said, before my time*'

William Wilkie Collins, *The Moonstone*

'*Live in danger. Build your cities on the slopes of Vesuvius.*'

Friedrich Nietzsche

CONTENTS

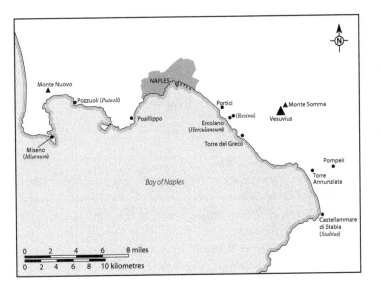

Map of the Bay of Naples showing Vesuvius and the major sites around it, both ancient and modern names given where relevant.

INTRODUCTION

Thunder rolls, smoke billows, flame spurts from the volcanic crater. Lightning zigzags in the dark. The dramatic effects are reflected, even enlarged, in the water below. But this is not the Bay of Naples but a featureless stretch of north eastern Germany. The Vesuvius that has sat in landscape gardens at Wörlitz, near Dessau, since the late eighteenth century will never erupt without warning.

An eruption is arranged every five years, twice on a single weekend, to impress the dignitaries from UNESCO, regional politicians and the locals. A splendid dinner is served on gondolas as the company is rowed through a watery maze of lakes and canals, under fancy little bridges from which the next course or a handful of rose petals come down. The destination is the 'Stein', the artificial island on which this Vesuvius sits. Out there are a small open air theatre (a miniature of the amphitheatres at Pompeii and Pozzuoli) and a little house, the Villa Hamilton. The creation of a prince made heady (and somewhat extravagant) by the infinite possibilities offered by the ambitious intellectual climate of the Enlightenment, the beguiling memories of his travels and the services of a wholly sympathetic architect, his Vesuvius was, and is, designed to flare up on a regular basis.

We've managed to buy tickets for Sunday 22 August 2010. But tonight, even the ingenuity of the portly Professor of Pyrotechnics at Cottbus University isn't equal to the elements. Thunder crashes come from every direction but the lurid greenish tinge in the night sky and flares of lightning are not of his making. The moon, such a key part of the effects the previous evening, is masked by cloud, then by cinematic sheets of rain. Storm and artifice play out their game, and storm wins. With difficulty each fully-laden gondola is forced to disembark and more than two hundred disappointed people disperse into the pitch darkness, like the fugitives from Pompeii. In the distance, the crater at the summit of the volcano still valiantly spews flame into the howling wind and driving rain, but the promised lava flow has been aborted and the full moon stays hidden.

Even in imitation, even under duress, a mountain with flame in its belly is thrilling. Judging from what they saw, or knew, the ancients believed that active volcanoes contained maddened giants confined by the gods to the nether regions and shifting angrily about. Equally, there was an anthropomorphic view, in which the volcano became a living, breathing organism, its effusions and ejaculations the result of circulating liquids and gases. Eruptions wracked the 'body' with spasmodic and violent reactions – a kind of geological epilepsy – the mountain spewing out matter and changing temperature at will, before settling back into long intervals of relative stability. Even the rational Sir William Hamilton, the elder statesman of Vesuvian studies, often referred to 'him' and 'his' ugly moods. From the seventeenth century onwards, people had been continually clambering up the flanks of the intermittently furious monster. They discovered that Vesuvius tended to reflect their mood and many found themselves transported by the site and, on occasion, by the volcanic performance too, just as we were that August in Germany.

In classical antiquity, despite or perhaps because of their familiarity with the principal locations, observers knew volcanoes to be unpredictable and uncontrollable, the repercussions unimaginable. (In that respect, very little has changed.) On the available evidence, the ancients made fair guesses about the causes of volcanic activity. They were fortunate to have had a deeply moving eyewitness account. Thirty years after the event, Pliny the Younger's letters to Tacitus recalled the cataclysm at Vesuvius in AD 79. In measured but vivid prose, he caught the last hours of his uncle, Pliny the Elder, as well as recalling his own experiences that day, the most cataclysmic event of his youth, indeed of his life. From the Renaissance onwards, Pliny's enthralling description served to remind travellers just how terrifying Vesuvius could be, a salutory warning as they set off in droves up the mountain.

A member of the Royal Academy of Sciences in Naples who witnessed the 1737 eruption of Vesuvius, commented that down the ages it had provided 'ample Matter for Reflection and Writing' while its inextinguishable fires and continuous new eruptions give 'modern Philosophers ... a sufficient subject to employ their Thoughts.' In the mid eighteenth century, Edmund Burke pointed out how extremes of scenery could prompt terror, a state of mind greatly to be desired in this context. Hardly had Burke developed his aesthetic theories when Vesuvius entered one of its longest active phases, during which sporadic eruptions continued for the best part of a century and a half. This book is about Vesuvius and that lengthy parade of men and women who found that the volcano mirrored their own moods and prompted conflicting, confusing feelings – among them fear, delight, curiosity or simply heightened emotion.

The dark symbolism of unpredictable violence and fearful transience lent itself effortlessly to political analogy. In troubled, introspective times, the volcano neatly suggested the overthrow of unwelcome regimes, the rise of revolutionary forces and popular movements or quite simply, the possibility of unforeseen change – for better or for worse. In Italy itself, Vesuvius was an image that helped formulate ideas of nationhood and, beyond that, to remind the world of the pivotal role that this small corner of the Mediterranean had played in its history.

Passive, Vesuvius was (and is) a set piece: a smooth, lavender-washed backdrop to chaotic Naples or the silent nemesis of the lost cities of Herculaneum and Pompeii. Active, even very slightly so, Vesuvius became a sensual, theatrical experience – playing pyrotechnic games against and with the moon or setting sun, reflections shimmering on the waters of the bay. The volcano had the capacity to violate anything in its path yet it provided innocent entertainment to audiences around the world – in amusement parks and dioramas, pleasure gardens and cinemas, in novels and even on the walls of art galleries and private houses. That cone, steep-sided because of the stickiness of viscous lava, looks very much like a firework, whose similarly cone-shaped packaging and evocative names tended to play up the volcanic link.

From the Renaissance onwards, the volcano (Vesuvius in particular) engaged the attention of alchemists and natural philosophers, Enlightenment thinkers, Romantic poets and artists, as well as those now known collectively as earth

The *ERUPTION of the MOUNTAIN. — or — The Horrors of the Bocca del Inferno;* *with the Head of the Protector SAINT JANUARIUS carried in procession by the Cardinal Archevesque of the Lazaroni.*

1. The eruption of Vesuvius in 1794, at the height of the French Revolution, was a gift to political caricaturists like James Gillray. Here he plays on the rituals surrounding the patron saint of Naples, San Gennaro (Saint Januarius), who was trusted to protect the city from disaster. In this case, the head is Charles James Fox's, borne by motley *sans culottes* (but with recognisable features) and the disaster has already occurred.

scientists. All of them chose what they wanted, or needed, to take from it. Vesuvius remains a profound and salutary reminder of what in life cannot be foreseen or brought to heel, as unpredictable as our own mortality.

The most famous and most accessible volcano in the world, Vesuvius came to reflect and symbolise the Romantic conflict, a *tabula rasa* for the unsettled psyche, while handily supplying orgiastic or at least sexual metaphors. Those who set explicit scenes by the crater include the Marquis de Sade and Mme de Staël. Those whose personal dramas were played out there include the Shelleys and Claire Clairmont and, at its foot, another tragic trio, Sir William Hamilton and his second wife, Emma, with her lover Admiral Horatio Nelson, often caricatured against the suggestive image of an ejaculating volcano. Visitors came in ever-growing numbers, in hordes, to see Vesuvius for themselves.

The volcano prompted self-examination and became a benchmark against which mood and the uncertainties of the soul were measurable: both Nietzsche and Freud linked Vesuvius to angst. The Surrealists were eager to exploit the possibilities of the metaphor, whether in film or on canvas. Volcanoes draw people and even, occasionally, immolate their disciples, the first being the early Greek philosopher, Empedocles. In June 1991 Katia and Maurice Krafft, French vulcanologists as eminent and experienced as any in the world, lost their lives at Mount Unzen in Japan as they were videoing it, felled by a pyroclastic flow, the deadly boiling gas bursting out without warning from a fresh vent.

Temperamental, atmospheric Vesuvius has been an obsession for many. Some stayed, watching it from close quarters for many years, attempting to unravel its mysteries. Others left, only to find the aftershock lingering long after they had returned to the placid landscapes of home.

The only active volcano on the European mainland, Vesuvius sits in a seductive setting, the keystone to every picturesque postcard view arching over the Bay of Naples. But since 1944, the most recent eruption, a demonic game of grandmother's footsteps has been going on, in which the population creeps ever higher and nearer to the old lava fields, seeming to tempt Vesuvius not to turn without warning and devour everything in its path. Against the astonishing reality of that last eruption, as seen in the black-and-white news clips readily available on YouTube, there's little to tempt visitors up now.

Much of the landscape is sour, derelict kiosks and cafés littering the lower region of the mountain with their telling detritus of failed or illegal tourist traps. Nothing, it seems, is ever removed; it all waits for another season's vegetation to wrap it up. Arriving by bus or car, tourists buy their tickets, pick their way up the final metres to the summit along an unlovely clinker path – marked by a new timber handrail already mostly adrift – and reaching the top, follow a guide along the track that hugs the crater's edge. Down below, nothing stirs but a minuscule wisp or two of vapour. Vesuvius can easily seem a huge anticlimax.

But a volcano, especially this one, is not innocuous. The dormant volcano is a fraudulent concept, dulling the responses of all except the teams of geologists and seismologists, watching and measuring (every hour of every day, they assure us), and who try to bring urgency to the discussion, as memories

2. Pompeii, with Vesuvius in the distance. A late nineteenth century commercial photograph taken by Giorgio Sommer's studio. German-born Sommer set up his business in Naples as early as 1856. He became enormously successful and specialised in art, archaeology and topography for a wide and growing market, the people who came in droves to visit Pompeii and Vesuvius, its nemesis. Conveniently, the photographer had the place to himself on this occasion.

of 1944 fade away, to inspire feasible evacuation plans and an infrastructure with which to carry them out. Writing in the aftermath of a global convulsion caused by a small subterranean volcano in the thinly populated interior of Iceland (to which some 70,000 people made their way, out of curiosity), it is all too clear, to me chillingly so, that the heavily built-up slopes of Vesuvius await a natural catastrophe on a quite unimaginable scale. The only uncertainty is when.

The following pages open a window onto a shared obsession that has endured for at least two millennia. Vesuvius has produced its own literature, imagery, scientific and universal insights but also, last but not least, it has engendered a huge amount of innocent delight and sheer astonishment – not just on Italian soil. I confess to becoming increasingly Vesuvius-struck during the writing of this book; I hope that my readers will follow suit.

1

....

EARLY DAYS

Perhaps the Earth's rumbling and fuming was less puzzling to the ancients than for us, so confident as we are of having the answers to everything and yet powerless to protect ourselves from violent natural phenomena such as active volcanoes or earthquakes. Our remote ancestors could shrug off their worries, since within polytheism volcanoes were the most obvious home of the many gods of fire and furnace. An eruption was, therefore, an expression of the gods' pent-up fury and so, while a worrying portent, not unexplained.

Since volcanoes were, from the viewpoint of the ancients, self-evidently caused by ever-burning fires at the heart of the earth, it followed that they would eventually burn out, their fuel spent. The only question was when that might happen. Recently, excavations have established that there was an exceptionally severe eruption of Vesuvius around 1600 BC at Avellino, just west of the current crater. The pyroclastic surge (also known as a *nuée ardente* – a burning cloud) must have reached Naples (as that of AD 79 did not). The footprints of people and animals fleeing their settlements as the jet of gas and volcanic matter, surging at a fantastic speed and at temperatures reaching 1,000 degrees centigrade, caught and killed them are poignant, if fragmentary, evidence of the deadly event. Logically, many other eruptions must have preceded it, of which no signs whatsoever remain. But now that we know something of the catastrophe at Avellino it overshadows all subsequent eruptions. It has, graphically, been described as the 'nightmare blueprint', forecasting that moment in the future when Vesuvius will reach, and overwhelm, the vast conurbation that is modern Naples.

Both the Greeks and Romans considered Vesuvius a sacred place from the evidence of its incredible fertility. The name is said to be a corruption of Vesouuios, son of Ves (Zeus), a figure close to the top of the hierarchy of deities. In Exodus, Moses received the Ten Commandments on Mount Sinai, 'all in *smoke* because the LORD descended upon it in *fire*'. Fire and brimstone were menacing and apocalyptical stage props in Judaeo-Christianity, whether summoned up at Sodom and Gomorrah in retribution for the people's heinous sins or within the Hell to which the Christian damned would be eternally confined.

In 1979 astonishing images taken from the Voyager space-craft revealed the nature of Io, one of Jupiter's several moons. The unprecedented scale and turbulence of this distant vol-canic landscape came as a shock to the scientific commu-nity, far surpassing anything on earth – both in the intensity of its heat and the sheer quantity of activity there. Io was a revelation. When space scientists began to name the most important of its nearly six hundred newly discovered active volcanoes, they turned to a thesaurus of deities of fire from every culture and corner of the world, both benevolent and malevolent. They included Prometheus and Amirai, his Georgian counterpart; Zamana the Babylonian god of sun and war; and Pele, Hawaian goddess of the volcano. Ancient

3. Aerial view of Vesuvius, showing the inert volcanic crater and, to the north, the half moon shaped sheer rim of Monte Somma, long extinct, rising behind. Vesuvius' long inactivity, it having been dormant since 1944, has lulled the local population into a dangerous sense of security.

myth and memory mediated between these thrilling discoveries within the planetary system.

Two millennia before those volcanic revelations from space, Vesuvius erupted. It would become the most famous volcano in the world. Two rocky peaks nestle: Monte Somma, the outer incomplete one, is a rocky palisade rising to 1,133 metres; Monte Vesuvio, the inner funnel-shaped cone, is a characteristic stratovolcano formed of highly viscous lava. The inner currently tops the outer by 150 metres (although every major eruption has changed its form and vital statistics). Divided by a barren valley, to the west the Atrio del Cavallo and to the east the evocatively named Valle dell'Inferno (Valley of Hell) the paired craters loom over Naples (just nine kilometres to the east) but only one is active. Monte Somma has been utterly dormant for the last 16,000 or more years. Vesuvius has been in temporary repose since 1944.

The origins and explanations for volcanic matter – fire, brimstone, noxious gases and other red-hot materials trapped underground – run through myth and scripture into early natural philosophy and beyond. By happy accident, most of the volcanic activity in the Mediterranean was easily accessible to the great thinkers of the Greek and Roman world and it is through their eyes, from surviving literature, that we see it. The sages of classical antiquity approached Vesuvius, Etna, Stromboli and other live sites, including the extensive Campi Phlegraei (Flaming Fields), with due respect for their mythic associations but above all with questing minds, acute vision and a dawning preference for the rational. The latter area, to the west of Naples, was a flat landscape, a sequence of volcanic calderas or shallow basins, some of which held water. Unknown to the ancients, the system extended further, under the Bay of Naples. In its way, that area was, and is, as actively volcanic as the famous mountains.

The most vivid clues to what lay underfoot came from the sulphurous wisps and scabrous crusts of the Solfatara (from the Latin, *sulpha terra*), the caldera which was the best known feature of the entire Campi Phlegraei. When he visited it in 1645, John Evelyn (recalling it later, but drawing heavily on several earlier accounts) used its nickname, the 'Court of Vulcan', and wrote of a place memorable for the heat ('almost unsufferable') and the 'variously colour'd Cinders' underfoot. It was 'no little adventure to approach them, however daily frequented both by Sick & Well, who receiving the fumes have been recover'd of diseases esteem'd incurable … a world of sulpure made'. In Christian times, this place was seen as a preview of Purgatory, its infernal qualities mediated by its curative powers.

The characteristics of a caldera (in effect a flattened volcano) such as the Lago d'Averno – a famously sinister, lifeless lake – are now known to indicate subsidence in the underlying magma reservoir – the yin to the yang of the volcano. When, very recently, astronomers gained more detailed images of Io, Jupiter's boiling moon turned out to bear a surprisingly strong resemblance to the Solfatara, with its scrofulous, sulphuric acid green-yellow surface.

Long before these discoveries Aristotle had posited that demonstrable and observable phenomena pointed to clear outcomes, and his empiricism led to a gradual devaluation of the status held by miscellaneous deities and the dismantling (or at least redirection) of primitive beliefs. The ancients sensed, and on occasion saw, that there was unseen turbulence below, a massive subterranean strife going on under the surface of the Earth. It helped to know that Etna was the home of the Cyclops who, blinded by Vulcan's fire, was blundering about in caves beneath the volcano. But returning to the demonstrable evidence, Virgil's description of Etna erupting in the *Aeneid* was as much an eyewitness account as any that came later. In the 1690s John Dryden conveyed it pithily – the 'pitchy clouds' rolling overhead and the 'flakes of mounting flames, that lick the sky' all manifest evidence of the 'fiery springs that boil below'. Etna had been far more rewarding to the volcano watchers of antiquity than the seemingly moribund Vesuvius.

Vesuvius is often referred to in the literature of classical antiquity. In one particularly seductive passage in his fourth epigram, the Spaniard Martial describes Bacchus and the Satyrs dancing in the vineyards and on the ridges above Vesuvius, in an area of glorious fecundity known as *Campania felix* (blessed Campania). So beautiful was this landscape, resulting from the rich minerals contained in volcanic deposits, that it offered, according to one translation, 'a retreat for which the gods of pleasure and gaiety forsook their most favoured abodes' – only for it all to be decimated at a stroke. 'All lies drowned in fire and melancholy ash; even the High Gods could have wished this had not been permitted them.' From now on Vesuvius is always pictured in dramatic, shocking contrast: before and after, fertile and infertile, paradisiacal and hellish – good strong polarities for a modern audience, too.

Martial drew the modern English poet Tony Harrison back to Vesuvius. It became the recurring subject of his poem 'The Grilling' (2002), in which he weaves together the ancient voices, along with seventeenth-century translations,

[15]

and Goethe and Tischbein his chosen Romantic companions, with whom he imagines enjoying fine wine and good conversation on the slopes of Vesuvius. One of Martial's translators was the Parliamentarian poet Thomas May, whose reputation had taken a dramatic downturn at the Restoration (though he was eventually reinterred in Poets' Corner in Westminster Abbey). The vicissitudes of his life endear him to Harrison, who finds resonance in May's choice of Martial. Thomas May's translation of Martial's epigram runs 'Where satyrs once in mirthful dances mov'd … Is now burnt downe, rak'd up in ashes sad. The gods are grill'd that such great power they had.' Fifty years later Joseph Addison tried again, with a far lighter touch; 'The frisking Satyrs on the summits danced:/ … Now piles of ashes, spreading all around/In undistinguish'd heaps, deform the ground: / The gods themselves the ruin'd seats bemoan / And blame the mischiefs that themselves have done.'

But even the ancients hadn't prepared Addison adequately for Vesuvius when he visited Naples in 1701–02. He confessed that until he saw it for himself, he had no idea of its impact. After all, the Roman poet Silius Italicus, whom he quoted, considered Vesuvius no more than a 'second Etna'. But for any politically engaged writer, powerful Vesuvius serves to illuminate dark places. Harrison, like others before him such as Shelley and Leopardi, uses the desolate volcano as 'a metaphor for all the fiery devastation of our times.'

Vesuvius' most resounding fame in antiquity rested upon the early history of Spartacus and his slave army. With fewer than a hundred armed fellow gladiator escapees, Spartacus took refuge in the dormant crater in 73 BC. The slaves had just broken out of their quarters in Capua, and with foresight

Spartacus directed them to head for Vesuvius. The outer rim of Monte Somma resembled a gigantic ruined keep with lookout towers and had just one entry point and thus, their pursuers guessed, a single exit. Spartacus's men could look down, unseen, upon the three thousand Roman infantrymen below, while busily manufacturing their means of escape: a quantity of improvised ropes and ladders made out of knotted vines and willow stems. With these, they clambered up the sheer walls of the crater and then shinned down the outer cliffs to take the massed soldiers, who were nonchalantly undefended, completely off guard. Spartacus's little force, their numbers now augmented by local shepherds and herdsmen 'all sturdy men and fast on their feet', routed Gaius Claudius Glaber's entire ragtag citizen army. Surprise had rendered them incapable of effective action.

The tale, as retold a couple of centuries later by Plutarch and Appian (it was the latter who identified Vesuvius as the precise site), gave the volcano a certain heroic reputation. The feared, ill-disciplined army that Spartacus soon amassed had begun, we believe, as an orderly and efficient military kernel which only survived thanks to their leader's prescient choice of hiding place. Peter Stothard, writing about Spartacus, judges the Roman engineer and bureaucrat Frontinus's account to be the most accurate since, ironically enough, he used the details for practical purposes to describe a site and a situation of particular awkwardness in the section of his military manual *Stratagems* that provides ideas and clever strategies for 'escape from difficult places'. More than a century after the event, he admired the clear thinking with which the handful of besieged men (Spartacus and seventy-four others, he believed) broke out of their high mountainous nest, leaving their horde of erstwhile captors askance. Later Frontinus became governor of Britain and took on the Welsh.

A detailed description of the characteristics of Vesuvius was to be provided by Vitruvius, the author of the first surviving architectural treatise and a far more renowned military engineer than Frontinus. Although the volcano was still dormant at the time he was writing (c.20 BC) the many trails of lava in the area gave him pause for thought. He admired the qualities of volcanic bi-products, without understanding their chemistry. Vitruvius describes a 'kind of powder' which when combined with lime and rubble becomes so solid and impregnable 'that neither the waves nor the force of water can dissolve them'. He is puzzled by its origins, though he knew of ancient tales about fires bursting out under Mount Vesuvius sending 'flames across the fields'. However, it was the high silica content in Vitruvius's wonderful 'sand' that gave the special mortar known as pozzolano, or *pulvis Puteolanus* (Puteoli was the Roman name for Pozzuoli), its impressive strength and waterproof properties. This bonding material could be used under water (for Ostia's docks) and for increasingly ambitious vaulted and curvilinear buildings. A revolutionary structure such as the dome of the Pantheon, with its massive unsupported span, entirely depended upon pozzolano, the key ingredient in 'Roman cement' and as important as any single element in the efficient and enduring construction of the monumental structures of the Roman Empire.

After the rediscovery of Vitruvius's work in the early fifteenth century and its subsequent adaptation into innumerable other manuals and treatises, engineers and architects gratefully embraced hydraulic cement, considering its properties near miraculous. Vitruvius came close to identifying the volcanic origin of stone in this area of Campania; and for architects such as John Soane, who came to Naples with his treatise in hand, Vesuvius was little more than a giant quarry for unusual materials. In fact, pumice (*lapis spongia*, sponge-like stone) and tufa (also known as tuff or tephra), compounded volcanic dust, were not particularly suitable for masonry. Paradoxically, from the Renaissance onwards, tufa (or, more often limestone which simulated it) would be considered ideal for grottos, fountains and other lightweight decorative garden structures, even for the Vesuvius that was built in Germany in the late eighteenth century. Other tougher igneous rocks formed out of magma reservoirs elsewhere, such as basalt and obsidian, were not identified as volcanic material for almost two millennia after Vitruvius.

Vitruvius also devoted considerable space in his treatise to the choice of suitable sites for particular types of building, including private residences. Many variants of the delicious *villa maritima* – with its airy vocabulary of garden terraces, arcades, colonnades, loggias and porticoes – were to evolve along the coast around Naples. It was an architecture developed to maximise the benefits of the natural beauty of the landscape, offering a choice of sea and mountain views, in contrast to the inward facing suburban *villa* hermetically sealed around an open colonnade, the peristyle. Some of these Roman waterside villas had jetties and reached down to the shore but most were terraced, stepping up several levels above sea level to better enjoy the prospect. *Amoenitas* (charm, delightfulness) was highly valued and life here emulated a civilised culture, the life of ease and fulfilment that the Romans believed had been a feature of earlier, Hellenic, times.

The sites of two large villas excavated at Stabiae (south-east of Castellammare) suggest that, exactly like their eighteenth-century successors outside Naples, they had been conceived to borrow views of the Bay in front, and Vesuvius to their rear. Wall paintings of similar villas show their relationship to the sea and, on occasion, to the hills behind. A difficult site, steep or confined, was an incentive – a way to engage with nature. Like those who began to build villas in great numbers following a Bourbon king settling in Naples in the 1730s, earlier generations of equally fortunate men, of Epicurean tendency, built here in order to drink in their delicious surroundings, just one strand within a completely sensuous existence that almost always teetered on the edge of vulgarity and extreme ostentation.

Others were looking ever harder at Vesuvius, the mountain, itself. Strabo, the Greek who, in the first century BC wrote a geographical account of the world as he knew it, has left the earliest surviving description of the volcano, pondering its very particular external characteristics. He noticed the dramatic contrast between that well-populated and intensively farmed land wrapping around the (by then long quiescent) mountain and its pale, flat, barren summit. There he observed 'pore-like cavities' in the blackened rocks and surmised that they had been 'eaten out by fire', suggesting that the entire mountaintop had once been alight, only to be quenched when the, to him, entirely mysterious fuel was spent. He considered volcanoes to be the earth's 'safety valves', the fire caused by subterranean hot winds or draughts fanning the rocks below, like sublime bellows, until they ignited, to seek an explosive exit via inadequate narrow passages to the surface of the earth.

This had been Aristotle's notion long before, and before him, that of Empedocles of Agrigentum in Sicily, the mid-fifth century BC Greek thinker, the first man to identify the four elements and an apparently prodigious figure, considered (on his own admission) to be a living god. His observations of a volcano, here Etna, gave rise to the potent and enduring idea of the earth as a kind of seething sponge, perforated by canals through which water and fire continually passed. The Sicilian volcano was reliably active and provided ready material for those who were struggling with theories of the earth. In contrast Vesuvius, as it then was, offered nothing more than a poetic, peaceful landscape surmounted by rocky outcrops.

This theory of volcanic activity was accepted whole-heartedly by Roman natural philosophers. Lucretius, in the sixth book of his *De Rerum Natura*, noted that the earth was 'amply fitted out/With windy caves'. Like all his contemporaries, he considered volcanoes to be hollow; 'The wind and air is everywhere around/ In these grottoes: the air becomes stirred up/ And turns to wind.' That air, by now warm, 'strikes up hot fire that rapidly/ Rises up high and hurls itself on out' by ejecting flaming material followed by billowing smoke and hurtling stones.

Seneca happened to be pondering the causes of earthquakes just as Campania was hit by a particularly violent seismic shock, which severely damaged both Pompeii and Herculaneum in AD 62 or 63. He could not know it, but he was grappling with the repercussions of a fatal conflict between shifting tectonic plates – an explanation which did not emerge until the mid twentieth century. That terrifying occurrence prompted him to pause and consider the psychological effects of natural disasters in the fourth volume of his *Naturales Quaestiones* (Natural Questions).

Even to a twenty-first century world audience, at risk of being numbed by a multi-media crescendo of terrifying natural cataclysms, Seneca's words are profoundly moving. How can anyone offer reassurance once all the certainties are lost, he wonders. 'Yet can anything seem adequately safe to anyone if the world itself is shaken, and its most solid parts collapse?' Worse still, 'What hiding-place do we look to, what help, if the earth itself is causing the ruin, if what protects us, upholds us, on which cities are built, which some speak of as a kind of foundation of the universe, separates and reels?'

⧖

By extraordinary good fortune in AD 79 the man in charge of the Roman fleet at the naval base at Misenum (modern Miseno), just across the bay from Neapolis (modern Naples), happened to be Pliny the Elder. Thus, Aristotle's heir, an astounding polymath, was on hand as Vesuvius erupted in a fashion never before recorded.

Pliny treasured every morsel of knowledge but was also committed to sharing his findings widely. He hardly slept, surviving on regular catnaps, and never moved without writing materials to hand and two secretaries, one reading to him, the other taking continual dictation (in winter with his hands encased in gloves). Somehow Pliny combined his diurnal official duties and attendance on the Emperor Vespasian with the nocturnal compilation of his multi-volume *Natural History*, the very first encyclopaedia. 'Nature, which is to say Life, is my subject', he wrote. Among the topics he covered was volcanism, though Vesuvius was not among the 'vents' he identified as being active. Discussing minerals, he identified certain types of sulphur which contained a 'powerful abundance of fire'.

In Rome, the annual festival of Volcanalia (or Vulcanalia) was held on 23 August and involved ritual sacrifice and celebratory games to 'allay wildfires, earthquakes and volcanoes' after the long weeks of searing summer heat. Volcanus or Vulcanus was the Roman god of destructive fire, whose shrine stood in the Forum Romanum. In AD 79 the festivities took place as usual, their objective to placate the forces of fire but, this time, with awful seeming prescience.

Pliny had already been intrigued, alert to such signs, by continuous recent earth tremors together with drying springs and broken water supplies. These hints of significant geological disturbance must have excited his curiosity hugely. His sister and seventeen-year-old nephew, the boy we know as Pliny the Younger, had been staying with him at Misenum. Recalling the first tremors, Pliny was to write that the tremors in themselves had not caused anyone great worry, 'because they are frequent in Campania'. But as he wrote to Tacitus thirty years later, one afternoon his mother suddenly alerted her brother 'to a cloud of unusual size and appearance'. On this day, probably 24 August, Pliny the Elder had taken a cold bath and had lunch before returning to work but he now urgently 'called for his shoes and climbed up to a place which would give him the best view'. For an avid natural philosopher who had never before witnessed a volcanic eruption, this was a thrilling, cathartic moment – the experience of a lifetime.

What they now saw, very far away, recalled his nephew,

was shaped 'like an umbrella pine, for it rose to a great height on a sort of trunk and then split off into branches, I imagine because it was thrust upwards by the first blast and then left unsupported as the pressure subsided'. At that distance, perhaps twenty-five kilometres as the crow flies (but more than sixty by land) the precise source of the immense, towering cloud was unclear. Pliny insisted that he must examine it more closely, and ordered himself a boat while asking his assistants to note 'each new movement and phase of the portent ... exactly as he observed them'. He offered to take his nephew with him on the water but, fortunately for posterity, young Pliny refused. He had a lot of work to do, including some tasks for his uncle.

As he was leaving the house Pliny received a desperate letter from a friend's wife, Rectina, beseeching him to organise help for those living in the area at the foot of the mountain, close to the shore. At a stroke his mission altered from one to feed his scientific curiosity into a full-scale naval rescue effort, involving all the warships he could muster. But as his boat approached the coast Pliny realised that they were already in the danger zone. 'Ashes were already falling, hotter and thicker as the ships drew near, followed by bits of pumice and blackened stones, charred and cracked by the flames: then suddenly they were in shallow water, and the shore was blocked by the debris from the mountain.' The helmsman advised Pliny to turn back but his master overruled him, insisting that they continue further south to Stabiae (southeast of modern Castellammare di Stabia), to the house of a friend, Pomponianus.

On arrival, Pliny feigned calm in front of his friends and servants and asked to take a bath. Afterwards they dined, watching Vesuvius ablaze with 'broad sheets of fire and leaping flame' given extra drama by the night sky. With authority, Pliny reassured everyone that these were merely either the bonfires of terrified peasants or their empty houses catching fire. Then he went to another room to sleep, 'his breathing was rather loud and heavy and could be heard by people coming and going outside his door.' Meanwhile the courtyard was filling with pumice and ash, to the extent that it would soon become impossible to open his door, and 'buildings were now shaking with violent shocks, and seemed to be swaying to and fro as if they were torn from their foundations.' They realised that they must all move again. Stabiae was less than five kilometres south of Pompeii.

To protect themselves from the rain of pumice, even though it was 'light and porous', the fleeing group covered their heads with pillows, tied on by cloths. Although it was daytime, 'they were still in darkness, blacker and denser than any ordinary night, which they relieved by lighting torches and various kinds of lamps.' Pliny suggested that they head to the beach. Arriving there, they found the sea churning, the boiling waves too wild to make boarding ship feasible. To their backs were flames and in the air an overwhelming, suffocating sulphurous smell. The group decided to run for their lives but Pliny, elderly and unfit, who had been lying on the beach on a sheet and drinking copious amounts of cold water (it has been suggested because of his heart condition), found it difficult to get up. 'He stood leaning on two slaves and then suddenly collapsed', as his nephew guessed, asphyxiated by the fumes (or felled by a heart attack). In either case, it was instant death for 'When daylight returned on the 26th two days after the last day he had seen – his body was found intact and uninjured, still fully clothed and looking more like sleep than death.' Pliny's nephew (and according to his will, his posthumously adopted son and namesake) recorded these final hours nearly thirty years later, in the two famous letters written at the request of the historian, Tacitus.

Pliny was the first in a long line of those who risked, or lost, their lives as they watched a volcano in action, mesmerised by the fantastic kaleidoscope of effects and fatally neglectful of their own safety. Long afterwards the monstrous umbrella pine form created by ejected materials was named after Pliny the Younger, on whose account our knowledge of the phenomenon is based. A Plinian eruption is among the deadliest of volcanic events although the mushroom cloud itself now evokes a far crueller weapon, the most devastating that nation can unleash on nation.

The middle-aged Pliny then recalled his own experiences. At Misenum they were now awake but stayed put. Pliny passed the night reading Livy's *History of Rome*. At dawn, daylight did not break through and buildings 'were already tottering'. People were behaving like sheep, following anyone who made a move and pushing hard from behind, the all-too predictable behaviour of a large crowd in the grip of panic. Next they discovered that the carriages 'began to run in different directions though the ground was quite level' and even wedging their wheels with stones was ineffective. Then, quite terrifyingly, the sea was 'sucked away and apparently forced back', the now-familiar effect of a tsunami. Out on the newly exposed sand 'quantities of sea creatures were left stranded'. In the other direction 'a fearful black cloud was rent by forked and quivering bursts of flame, and parted to reveal great tongues of fire, like flashes of lightning magnified.' Pliny's

4. Angelica Kauffman's *Pliny the Younger and his mother at Misenum* (1785)
dramatises the moment Pliny's mother, sister of Pliny the Elder, tries to
persuade her teenaged son to save his own life and leave Misenum while
Vesuvius continues to wreak havoc across the Bay of Naples. He refused to
go and long after wrote his famous letters to Tacitus vividly describing those
terrible days in AD 79.

mother pleaded with him to go, to save his own life and not let her delay him, she being so much older and slower. He ignored her words, grabbed her hand and pulled her along the road after him.

Although it was daytime, already 'darkness fell, not the dark of a moonless or cloudy night, but as if the lamp had been put out in a closed room.' A claustrophobic black vapour now enveloped them and threatened to choke everyone. He remembered the wave of shrieks, shouts and wails, 'some were calling their parents, others their children or their wives, trying to recognise them by their voices.' Some wept, others prayed for death. 'Many besought the aid of the gods, but still more imagined there were no gods left, and that the universe was plunged into eternal darkness for evermore.' Every kind of rumour ran through the fleeing crowd, words as incendiary as the lava fires running down Vesuvius.

Every so often, they stood up and shook off the build-up of ash lest it weigh them down. By now Pliny was convinced that this was likely to be the end of life, his own and possibly of life in general. Then at last the sky slowly lightened as the sun began to shine faintly through, like a partial eclipse. Everything was revealed covered with a drift of thick white ash, as if after a heavy fall of snow. They had not gone far and decided to turn back to Misenum and Pliny the Elder's house. There, as the ground continued to quake, they awaited the fatal news.

The ash had fallen as far away as North Africa and was disastrous for farmers throughout southern Italy. The eruption, as revealed by examination of the depth and layers of deposit, continued for some time. The unfolding drama is revealed, with chilling immediacy, in the condition of the bodies of the fugitives from Pompeii. The first victims are encased in pumice. The further they fled, the thicker was the coating. Those who reached the eastern city gate and beyond, who were already on the road which led out to the countryside, bordered by the town necropolis, must have thought they were heading for safety. But they were hit by the irresistible and sinister pyroclastic surge, a geological afterthought brooking no obstacles or evasion. A hurtling jet of gas, carrying within it the recent detritus of volcanic eruption at immense speed and horrifyingly high temperatures, it quite simply incinerated everything in its path. The surge was, in effect, a horizontal H-bomb.

At Herculaneum, the pyroclastic surge wiped out the population of the town at a stroke. The victims were summarily executed, their bodies compacted (and obliterated) under some twenty-four metres of volcanic material. Here, far closer to the volcano, once the thrust of the gas surge had turned in their direction, people had no chance of survival.

The deadly effects of this latter episode have been disinterred from an accumulation of evidence and can be compared to the pyroclastic cataclysm that accompanied the 1902 eruption of Mount Pelée on the Caribbean island of Martinique from which, in the principal town Saint Pierre, there were but two survivors out of a population of 30,000. At Vesuvius, the archaeological evidence, seen in the light of increasingly sophisticated geological understanding, points to an episode even more terrifying than had been previously realised. Those two crucial poetic letters from Pliny the Younger, written thirty years after the event, offer a snapshot – a single haunting episode in a prolonged nightmare.

Yet the level of physical destruction found at Pompeii has

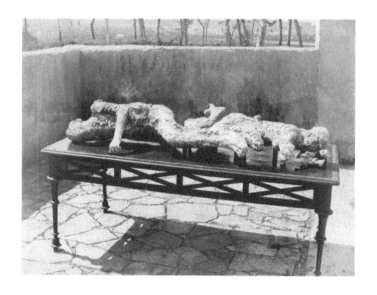

5. Casts of two victims of Pompeii, made by the archaeologist Giovanni Fiorelli in the early 1860s. They lie incongruously in a museum display cabinet, outdoors. The poignancy of these figures in their death throes caught the public imagination and soon became the major popular attraction at the site.

confused matters. The severe earthquake of seventeen years before (the event that prompted Seneca's poignant musings on human reaction to natural catastrophe) had probably left many buildings, both public and private, in ruins. Thus, some of the wreckage excavated at Pompeii is likely to predate the eruption; other wreckage may be the result of later looting, pillaging and recovery of property – those all too human and desperate responses after disaster.

There has been confusion, too, about the human cost. The majority of the Pompeian population apparently made it to safety. Some had heeded the seismic forewarnings, though whether long before or merely in the immediately preceding hours or days, is unknown. The apparent distance of Pompeii from the mountain, some twelve kilometres, may have made the population fatalistic about the risk. Almost twice as many of the dead were victims of the pyroclastic flow rather than from the initial showers of pumice. But it is these, their bodies re-formed by a kind of Victorian injection moulding, who seized the popular imagination from the 1860s onwards, such apparently lifelike and legible cadavers.

As it became clear that Pompeii, Herculaneum and other settlements had been razed, there must have been enormous disillusion in the efficacy of the gods. Reciprocity lay at the root of the Romans' dealings with the deities, whether the powerful Roman gods or the village idols, the little figures of local tradition. Properly honoured, they would assure that no harm came to the population, but failure to attend to those obligations the gods expected was to tempt fate. Following this catastrophe the Emperor Titus seems to have behaved well – given that the gods appeared to have failed the local people. He organised prompt relief for the area and came in person the following year – an act of charity celebrated by Mozart who, as a boy, visited Pompeii and played music within earshot and sight of the active volcano, in his final work *La Clemenza di Tito*.

Ash fell as far away as Constantinople when Vesuvius next erupted in AD 472, but as the volcano fell quiet so too did the discourse around natural phenomena. Haraldur Sigurdsson refers to the following centuries as being wrapped in 'scholarly amnesia', a comfort blanket of religious conviction that ordained that all was God's work, and that He knew best. The introverted and intense stance of early Christians – asceticism – had elevated their certainties and kept them well away from idle speculation.

2

....

MIRACLE OR SCIENCE?

In 37 BC Horace recorded an ancient pagan rite in southern Italy. He wrote, in *Satire 1.5*, of Gnatia in Apuglia, a 'town built amidst troubled water', where the people of the place encouraged him to believe that 'incense placed on the sacred threshold liquefied without flame'. Horace knew the rules, as laid out by Epicurus and Lucretius: everything could be explained by reference either to natural causes or the laws of science. With their authority and a comfortably empirical stance, he confidently distanced himself from the mumbo-jumbo. But for the people of Gnatia, as for the medieval Christians in Naples, the transformation of a dry substance to liquid by the agency of divine intervention was credible.

The patron saint of Naples is a late-third-century bishop of Benevento, named Januarius, a victim of Emperor Diocletian's energetic persecution of Christians. After he was revealed to have given the holy sacrament to an imprisoned colleague, he was offered to the lions. But Saint Januarius, as he became (San Gennaro to the Neapolitans) pacified the beasts and saved his life, only to be subsequently beheaded near Pozzuoli. However, a Christian woman presciently saved a phial of her bishop's blood. By time Emperor Constantine ceased the persecution of Christians, the martyred bishop's skeleton had been lost in a catacomb. Once found, centuries later, San Gennaro's remains were accorded suitable prominence. In 1304, Charles II of Anjou commissioned a silver reliquary for his head. Together with other bones and two phials believed to contain his dried blood, San Gennaro was reinterred in the new cathedral, and a chapel dedicated to him and his miraculous attributes.

The first recorded ritual based around the miraculous liquefaction of his blood was as late as 1389. From then on, San Gennaro was entrusted with the arduous task of protecting Naples from disaster, be it the eruption of the neighbouring volcano, an earthquake or the outbreak of plague. The protracted ceremonies centred upon the transformation of the murky sludge in the thick glass phials to liquid. In medieval times these took place twice a year – each time extending over a week – once in early May and again beginning on 19 September, the saint's name day. With the obvious link to the sacrament of the eucharist, the miraculous liquefaction seemed to fall reassuringly within the ambit of Christian liturgy. In time, a growing number of similar occurrences were recorded in and around Naples on various obscure saints' days but more reliably on those of St John the Baptist, St Stephen and St Pantaleone. When San Gennaro's blood liquefied in the Duomo it was said that a stain on a block of basalt at Pozzuoli, where the martyr bishop had finally met his death, appeared to glow red. A belief in the truly miraculous was, after all, the only sure protection against the unknown and unknowable.

Christianity brought strong reaction to the rational approach advocated by natural philosophers. Why, asked the early believers, should a dramatic episode such as a volcanic eruption *not* be God's work, an event dictated by and even foreseen in the Bible? The fires of the Earth were obviously linked to the fires of Hell or, at the very least, that uncomfortable antechamber, Purgatory. In the early centuries of the Christian era, scholarly monks mounted a close watch on active volcanoes, if for no better reason than to prove conclusively that they were the work, and residence, of the devil. In AD 968 an eruption of Vesuvius was recorded, followed by several more at close intervals. Since the end of the world was expected in AD 1000, those reports were slightly suspect. Vesuvius, as always, was at the service of numerous interests. Nevertheless after June 1139, the date of a considerable and well-recorded eruption, Vesuvius does seem to have fallen dormant for several centuries.

In the Renaissance, those level-headed empirical accounts from antiquity which read as freshly as if they had been written yesterday, became widely available both in Latin and in translation. Volumes devoted to 'natural history' – in reality any and every aspect of science and topography – poured off the printing presses of Antwerp, Amsterdam, Paris, Venice and London as timely reminders that the unexplained spasms of the earth were matters of historical record and deserved closer examination.

Anyone travelling south from the 1500s onwards was consciously treading in the footsteps of the ancients. The well-lettered knew that Pliny the Elder, the polymath father of natural history, had risked and lost his life while observing an eruption of Vesuvius, having read his nephew's letters to Tacitus which retold the events of late August AD 79. They were essential reading for those who were planning to take the plunge and go as far south as Naples. Prints and maps

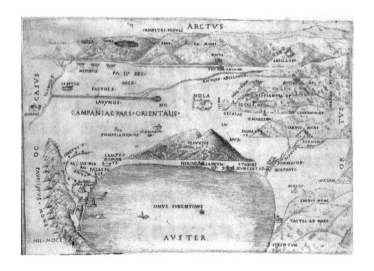

6. Girolamo Mocetto's engraved map of the Bay of Naples from 1514 shows
Vesuvius (and unmarked alongside it Monte Somma) in relation to the
still undiscovered ancient settlements that had been buried by the volcano
in AD 79. The sites of Herculaneum and Stabiae are correctly marked but,
disconcertingly, Pompeii has moved to the north west of Naples.

were equally beguiling evidence. Girolamo Mocetto's etched bird's eye view of the Bay of Naples (1514) speculatively sited the recorded, but still undiscovered, towns of Herculaneum, Pompeii and Stabiae at intervals around the edges of a neat oval basin, more birdbath than natural harbour, over the very centre of which loomed twin-peaked Vesuvius.

Young Thomas Hoby, best known as the translator of Baldassare Castiglione's *Book of the Courtier*, was an early British visitor to Sicily and the remoter regions of southern Italy in 1550, by which time the region had already been in the fierce grip of a succession of Spanish Viceroys for half a century. The twenty-year-old knew his Pliny well and was alert to and appreciative of his surroundings. Around Naples he enjoyed the mineral springs, as good for health as for 'pleasure and disport' while noting signs of subterranean activity, 'hills of sulphure, upon whiche are sundrie little holes that with great force cast owt verie hot smoke and sum flames of brimstone.' Alongside that, the astonishing fertility of the fabulous *Campania felix* impressed him greatly, mirroring the idyllic and pastoral setting for shepherd-poets conjured up in *Arcadia* (1504), by Jacopo Sannazaro. A Neapolitan, he had also presciently included a 'vision' of the lost and buried towns.

Thomas Hoby, reaching Vesuvius, was struck forcibly by its exceptionally fertile foothills and the harshness of the higher ground, especially 'the toppe, whiche is full of great sharpe burnt stones'. He continued: 'This hill burnt in Plinie's time, who went abowt to serche owt the cause of yt and was smodred in the smoke.' But he reserved his greatest excitement for Monte Nuovo (literally the 'New Mountain'), near Pozzuoli, which he persistently referred to as Tripergola.

This little peak 'arose sodainlie owt of the plaine sandie ground upon St Michael's day in the yere of owr lord 1538 with suche a terrible noise and such violence that it cast stones as far as Naples.' The eruption had terrorised the city and ash had fallen up to twenty-four miles distant. Monte Nuovo had continued to burn for some time but was dormant by the time he arrived. A newly minted mountain was an unsettling, mystifying sight. When Mocetto etched his view, little more than twenty years earlier, it had simply not existed. But now, gratifyingly, Thomas Hoby could lay claim to his very own Renaissance volcano.

A race began to find the explanation for the convulsions that had suddenly begun to shake this part of the Mediterranean. In the early seventeenth century two major eruptions, first at Vesuvius, then at Etna (which rarely acted in such convenient concert), allowed curious travellers the opportunity to peer into the core of the earth and try to surmise what was occurring beneath their feet. In 1693 came a fearful earthquake in the south of Sicily. Clearly, the very intestines of the planet were churning; in a vividly anthropomorphic image, the early seventeenth-century German astronomer Johannes Kepler had proposed that volcanic eruptions were the tears and excrement of the Earth.

Natural philosophers, the scientists of their age, were determined investigators, following those inspiring men of classical antiquity. In April 1626 Francis Bacon wrote to the Earl of Arundel to explain why he had taken refuge in his friend's Highgate mansion. (He died there, soon after.) Airily, he compared himself to 'Caius Plinius the elder, who lost his life by trying an experiment about the burning of the mountain Vesuvius.' His own experiment involved 'the conservation and induration of bodies'. John Aubrey's picturesque version of the event involved Bacon buying a chicken and stuffing it with snow to demonstrate what an efficient preservative ice was. In fact, he may have been experimenting with self-medication, rather than frozen chicken. If he did die from a fatal overdose of saltpetre or opiates, rather than from a pre-existing condition, he had subtly converted his misuse of dangerous substances into a more heroic claim to be emulating Pliny, the man who incurred death while engaged in scientific observation.

The motto chosen by the Royal Society soon after its foundation in 1660 came from Horace, *nullius in verba*, 'take no man's word for it'. It was no empty phrase. Natural phenomena had to be observed and recorded with scrupulous care if anything at all was to be learned from them. From the start, travellers, Fellows and friends of Fellows all flooded the learned society with their communications. Eyewitness accounts were best and they all watched Etna and Vesuvius like hungry hawks.

Volcanoes provided irrefutable physical evidence of geological activity. The interaction of the wind, water, heat and chemical subterranean reaction believed to be responsible for their activity, must be examined. Sometimes their enquiries pointed down a blind alley. Since the southern Italian sites of volcanic activity were all close to the sea, a link was deduced between salt water and the unimaginable heat far below. Some form of internal combustion was common to every theory, even though the image of an everlasting fiery furnace at the core of the earth had been challenged as early as 1632 by an explanation pointing to chemical reactions between materials at modest depth. Equally, the importance of sulphur was not borne out by the evidence (molten material from Etna had no sulphurous smell) and the chemical composition of lava appeared to be closer to that of glass, a form of silica. Although they often nudged close to the answers, drawing accurate evidence-based conclusions from the close study of rock, ash, minerals and lava deposits, the near-truths reached by several early earth scientists remained largely overlooked.

⧗

Vesuvius had long been quiescent. Before the eruption of 1500 it had not stirred since 1139. Then on 16 December 1631 it erupted violently – a terrifying event for which no one had been prepared, for nobody alive had experienced the volcano in action. Within days an immense cloud form rose, reminiscent of the local umbrella pine, reaching many kilometres into the sky. Almost simultaneously an earthquake sucked the sea inland by several hundred metres; floodwater and mud slips, mistaken for lava flows, compounded the horror. Once more the physical relationship between the fixed outline of the dead Monte Somma and the volatile Mount Vesuvius, now all too alive, changed significantly when the inner cone collapsed. The delicious, Arcadian landscape of Sannazaro's vision – the pastures, chestnuts and vines – was utterly obliterated and replaced by a mysterious level 'sandy' surface.

An estimated four thousand people died and some thirteen municipalities were laid to ruin. Those who had read Pliny the Younger's account must have had a sense of déjà vu, since the pattern of events was so similar to the eruption of AD 79. The horror of what unfolded consisted of both predictable and unpredictable elements. Scholars, many of them monks, recorded it for posterity.

The eruption of 16 December 1631 was the first since AD 79 to involve a pyroclastic flow. So sudden and deadly was it, that the authorities, both civil (in the person of the Spanish Habsburg Viceroy, Count Monterrey) and religious, took cover behind their Counter Reformation reflexes and the miraculous powers of the patron saint of Naples, San Gennaro.

The degree of devastation to the towns and villages, vineyards, fields and livestock all around the eastern flank of the volcano was, for now, explained as a divine punishment for the awful misdeeds of the Neapolitans and their neighbours. Those from outside the city were barred from entry (and thus safety) for fear of them carrying the plague. Beyond the city boundaries other less omnipotent and certainly less miraculous saints were vilified for their failure to provide adequate protection. The episode revealed an enduring connection with the beliefs and practices of ancient times, a visceral memory of the attributes of particular gods. The empirical reading of volcanic phenomena was under siege once more from superstition, atavism and the miraculous.

The eruption of 1631 fuelled a resurgence of printing and publishing in Naples, producing more than two hundred narratives which reinforced the political status quo by pretending order in the face of chaos, celebrating the efficacy of the miraculous saint or providing distraction in the form of planetary and astrological readings or spurious philosophical, religious or scientific insights. An avalanche of prints, from crude woodcuts to fine engravings, commemorated the event, often with San Gennaro or the Virgin and a crowd of protective saints emerging, somewhat smugly, from the suddenly comforting clouds overhead. None of this contributed to a better understanding of the volcanic process; it was as if natural philosophy, like those who clambered up the ash and rubble-covered flanks of the volcano in quieter times, was taking two steps back for every one forward.

That December, the religious orders had gathered up their relics, in particular those associated with San Gennaro, and processed from their quarters towards the erring volcano. As the hours passed, the files of penitents grew ever longer and their actions ever more masochistic. Self-flagellating, they tramped the city streets barefoot and bleeding. The distraction of these officially orchestrated and collectively hysterical events did nothing to help those who were suffering beyond the city boundaries. As Haraldur Sigurdsson writes, invoking San Gennaro and his angelic helpers as a 'fire brigade' (Filippo d'Angelo's drawing shows them pouring pails of water down from the clouds), merely served to stoke the panic. Those with the means had already fled the area and others were given negligible practical help, merely the placebo of spurious devotional proceedings, generally orchestrated to take place when the worst was over – to prove their efficacy. Even the astrological charts were adjusted to show their prescience, after the event.

Since the trail of destruction stopped just short of the city, at the Maddalena Bridge, San Gennaro was considered to have intervened to excellent effect. Soon after the catastrophe, a decorative scheme was commissioned for the royal chapel at the cathedral. A number of leading Baroque artists, including Jusepe de Ribera and Domenico Zampieri (a contentious choice since Domenichino, as he was known, was not a Neapolitan but a Roman painter), elaborated on the theme of the life and achievements of the saintly bishop. The cycle of oil paintings on copper and a series of frescoes covered every inch of plaster on the walls and wrapped the surfaces of the cupola above. They showed his martyrdom and trumpeted the protection he had given the city from pestilence, marauders or acts of nature, and most particularly his care for Naples during the recent eruption.

From now on the ceremonies associated with San Gennaro became ever more elaborate and frenetically Catholic. Protestant visitors to Naples were bemused by their experience of the arcane rituals surrounding the miraculous powers of the martyred Roman bishop. San Gennaro's role was to placate the black forces of destruction, the wages of sin, and keep the city unharmed. Fiery apocalypse, divine judgement crashing down upon the sinners at the foot of the mountain, sat within Counter-Reformation notions of chastisement and retribution as did the more anthropocentric view, in which the erring populace had only itself to blame for the situation. Whichever way they looked, Neapolitans were subject to awful afflictions, hazard and punishment.

In this Baroque capital of religious ritual, in which hardly a week passed without processions, pyrotechnic displays or intense prayer to any one of an entire calendar of saints, the anniversary of the eruption, 16 December, was added to San Gennaro's personal diary. After his relics had wended their way through the cavernous streets to the cathedral, accompanied by dignitaries and a vast congregation, they now returned to a chapel wholly given over to the cult of the saint. San Gennaro's head, encased in an elaborate reliquary, was placed by the altar along with the phials of his blood, which a series of priests rocked slowly to and fro for hours, if not days. A Greek chorus, his female 'descendants', provided a crescendo of prayer and keening until the dusty substance turned to liquid (or, should it fail, dissolved into hysterics on cue). The rest of the city heard the news of a successful liquefaction, the awaited and fully expected miracle, when the cathedral bells peeled out while, still inside, the congregation burst into the *Te Deum*. Anyone who could reach it smothered the reliquary with kisses. In the seventeenth century the line between theatrical spectacle and religious ritual became increasingly blurred. Like the *presepi*, the traditional Neapolitan nativity scenes of church and street, these diverting rituals, with their thick Christian varnish coating the slight atavistic remnants, served the authorities well as distraction from harsh reality and incipient disorder.

John Milton travelled to Naples in 1638. In *Paradise Lost* his Satan springs out of a pyramid of fire, surely singed by volcanic debris, a vision of Prometheus. The Palace of Pandemonium is built on the side of a volcano. Even though Milton's device was metaphor and his motive was moral, Vesuvius served his purpose well and it has been suggested that he was prompted by the descriptions of the event from his host, Giovanni Battista Manso, the poet, writer and patron. Manso wrote a letter, as he watched the eruption from Naples, showing a less visually startling, more sinister reality. Down the mountain came a river of material, 'between flames and smoke', for the pyroclastic flow, a hurtling, burning blast of gas, had masked the characteristic lightning and flames, which still burst through from time to time. (The 1631 eruption ejected very little lava, so that there was little sign of the usual slow rivers of molten material characteristic of volcanic events.) Manso employed a military metaphor to convey the intensity of the noise, comparing it to an artillery bombardment during which the ground shook continuously. To Milton, a man more familiar with the mundane, unchanging landscape of southern England, Manso's firsthand description of those extraordinary, turbulent days must have been a revelation.

The synergy between natural phenomena and unearthly powers was evoked, very literally, by the Neapolitan born painter Salvator Rosa who depicted Empedocles, the fifth-century BC Greek philosopher and self-styled divine, immolating himself within the crater of Mount Etna. In Rosa's preparatory drawing, even more explicitly than in the finished painting, he is shown plunging down, his robes billowing out, from a ledge on the sheer rock face into the void below. With this leap he would prove his immortality to his followers. (Rosa included the shoe that he left behind, giving the lie to Empedocles' claims). Milton wrote of him as 'He who to be deemed/ A god, leaped fondly into Etna flames'. That startling image, whether conjured up in paint or in lines on the page, evoked man's terrified fascination with the implacable aspects of nature in the face of our certain mortality far more vividly than the cycle of rousing but essentially well-rehearsed scenes that surrounded San Gennaro's acolytes when they knelt to pray in his chapel.

Once the news of the 1631 eruption, so horrifying and yet so exciting, reached northern Europe – and now that England and Spain were at peace – travellers were tempted to venture to Naples and take an excursion to Vesuvius and the surroundings, now a diabolical post-eruptive landscape. In his 1670 guide to Italy, which includes the first use of the term 'Grand Tour', Richard Lassels suggested that hiring a horse for two days would be quite long enough to visit both Vesuvius and Pozzuoli, with its amphitheatre, temple and remnants of a Roman bridge, at the quite moderate cost of around thirteen to fourteen crowns per person for the round trip. Despite occasional finds, the archaeological riches at the mountain's foot were unsuspected and the buried cities would not be added to the itinerary, or more time allowed for visits to the newly discovered sites, for more than a century.

The seventeenth century is littered with men whose intellectual ambition and energy mirrored that of Pliny the Elder – a continual inspiration to his admirers. The French thinker, Peiresc, visited Vesuvius in 1601 as a mark of homage to the great encyclopaedist. In turn, his German Jesuit admirer and follower, Athanasius Kircher, who was to become the resident polymath at the Collegio Romano from 1635 onwards, celebrated Plato's desire 'to know everything' – an ambition which he fulfilled both by prolific collecting and tireless publishing. But first he travelled.

In 1638 Kircher spent some months in Malta and Sicily as the father confessor to an aspiring German cardinal. During this tour, Kircher visited Mount Etna and experienced one of its regular eruptions. Back on the mainland, he then turned to Vesuvius, which bore clear evidence of the far more exceptional recent violence and was still intermittently active. To examine it better, he arranged to be lowered deep into the crater in a basket. This daredevil experience, never found necessary by later natural philosophers, took him close enough to death – 'methoughts I beheld the habitation of Hell' he claimed – to alter the entire course of his work. He embarked upon a lifelong investigation of nature's secrets, though steadfastly keeping the reference points of his faith in view.

Kircher's engravings, dated 1638, eventually appeared in the long delayed *Mundus subterraneus* (Subterranean World), the earliest volumes being published simultaneously in 1665 in (Protestant) Amsterdam and (Catholic) Rome. In 1669 an opportunistic English printer selected almost seventy pages of the text, and one plate, and published it as *The vulcano's, or, Burning and fire-vomiting mountains, famous in the world, with their remarkables.* The preamble explained that he felt a selection of Kircher's text should be 'exposed to more general view in English: upon the relation of the late wonderful and prodigious eruptions of AEtna, thereby to occasion greater admirations of the wonders of nature (and of the God of nature) in the mighty element of fire'. Curiosity regarding volcanic activity was growing. In the same year the earl of Winchilsea published an illustrated account of the eruption of Etna. The Sicilian volcano could be depended upon for almost continuous action, while Vesuvius, so often slumbering, never would be.

Even though Kircher was constrained, as a Jesuit, to emphasise divine purpose and to remain within a strict biblical chronology of some 6,500 years, his own investigations suggested other conclusions and great uncertainties. Stephen Jay Gould has written of his admiration for the 'quality of his doubts' and for Kircher's resolve as he struggled with material of which, he admitted, he had only very limited understanding. On one occasion at least, Kircher provided a sound explanation for an apparently miraculous event. In 1660 tiny crosses (immediately construed as crucifixes) appeared on

Neapolitans' clothing following an eruption. Kircher wrote a pamphlet about these 'prodigious crosses' explaining that they were in fact the result of a particular chemical response, in which pyroxene crystals combined with volcanic ash and then spread through the warp and weft of the linen threads. (This theory makes perfectly good sense to modern geologists.)

The spectacular jumble of objects that Kircher was accumulating at his rooms in the Jesuit College in Rome (and which later became, properly speaking, a museum) reflected his encyclopaedic, though sometimes fallacious, mind. The eccentric Jesuit and his bizarre collection drew early Grand Tourists to his door. In droves they hurried there, to be admitted upstairs by an elaborate version of the entryphone, a speaking tube of Kircher's own devising. Among these bemused admirers was John Evelyn who in November 1644 found himself in a life-size cabinet of curiosities. He described Kircher's study as a paradise of 'perpetual motions, Catoptrics, Magnetical experiments, Modells, and a thousand other crotchets & Devises'. For a young man who had already chosen as his motto *omnia explorate, meliora retinete* (explore everything, keep the best) this was very heaven.

Kircher himself usually acted as guide and, no doubt, introduced whatever topic was currently engrossing him. His ability to 'read' Egyptian hieroglyphics, which so impressed his visitors, including Evelyn, may have turned out to be somewhat exaggerated but his voracious appetite for universal knowledge was sympathetic to the intellectually curious, who found him a questing mind in the best tradition of Pliny. Kircher illustrated volcanic phenomena by enacting small-scale eruptions in his room, using a candle behind red waxed paper to convey the effects of molten lava on the

7. Athanasius Kircher's engraving of a section through Vesuvius, made
in 1638, owes much to imagination and something to observation. A
polymath, Kircher claimed to have been lowered into the crater in a basket
to experience a recently active volcano at close quarters. In this print he
effectively contrasts the interior, with the molten core far below, with the
charming orchards and fields that wrap around the exterior.

mountainside. His visitors were both enthralled and alarmed.

Kircher's own engravings showed an amalgam of what he had seen and what he guessed to be there, imaginatively and graphically combining section, elevation and perspective, the volcanic cones dissected both horizontally and vertically. One plate of 'Vesuvius' illustrated a complex system of flaming subterranean cells linked by arteries to a perpetually burning heart. Pressure from these underground systems, he held, led to the sporadic and unpredictable actions of volcanoes. In his cross-section of the earth Kircher added melon-cheeked cherubs in the corners to show the four winds, the usual convention for terrestrial maps. Yet his fiery labyrinth, off which a complex of tunnels led to volcanic outbreaks at intervals around the world, had a certain intuitive accuracy about it. Kircher imagined reservoirs of both fire and water at the heart of the earth. Steeped as he was in the ideas of Empedocles, he was simply illustrating the 'sponge theory' of the earth, an appealing notion that the Greek philosopher and close observer of Etna had put forward in the fifth century BC and which was not to be entirely discredited until the nineteenth century.

From the late 1660s, as Kircher's volumes upon natural phenomena such as earthquakes and volcanic eruption became available around Europe, he was to become the most influential of authorities on the subject while his Jesuit credentials rendered him entirely theologically sound. His dedication, to Pope Alexander VII, referred to a world 'drawn forth from nothingness through the Chaotic Mass'.

Kircher's plates, though they were not published for another thirty years, may have been available to visitors such as young John Evelyn in the 1640s. It was, most probably, Kircher who prompted him to go and see the pyrotechnics of Vesuvius for himself. He set out for the south of Italy in early February 1645, 'the Non ultra' of travel as he put it since he would soon be treading the very soil of Classical antiquity. As if to prepare travellers for what lay ahead of them, the Via Appia was lined, Evelyn wrote, with 'antient Sepulchers, Inscriptions and ruines'. It all amounted to a kind of anteroom, a lobby to the Roman world.

A week after arriving in Naples their party, which included his friend Thomas Henshaw, set out for Vesuvius. They approached in the usual way, crossing the bridge that marked the outer limit of the 1631 lava flows, where San Gennaro's miraculous intervention had supposedly taken place. Once they had left their mules behind, they were reduced to

> *crawling up the rest of the proclivity, with extraordinary difficulty, now with our feete & hands, not without many untoward slipps, which did much bruise us on the various colour'd Cinders wth which the whole Mountaine is cover'd, some like pitch, others full of perfect brimstone, others metalique, interspers'd with innumerable Pumices (of all which I made a collection). We at last gained the summit … it presents us one of the goodliest prospects in the World.*

For Evelyn, Kircher's avid disciple, the richness of his booty was more than adequate compensation for any superficial wounds.

Although Evelyn's diary was largely composed, and

amplified, long after the events in question, his entries for the visit to the volcano convey a heady excitement, suggesting that they were written on the spot. He may also have felt a more than usually close interest in what he was seeing; the Evelyn family fortunes had been built on a gunpowder monopoly and he was already taking a particular interest in the natural sciences, modern chemistry and geology. The supposed link between volcanic eruption and gunpowder appeared to lie in the strongly sulphurous smell of the former and the fact that sulphur was an ingredient in gunpowder. This neat conjunction intrigued the natural scientists of the time, questing alchemists who modified and codified their evidence as they grew into thoughtful (and sometimes sceptical) chemists. Like John Evelyn, Thomas Henshaw, both of whose parents were described as 'chemists', would become a very early Fellow of the Royal Society.

For Evelyn the experience of reaching the very rim of the crater touched all the senses. First of all, like every visitor who has climbed up there, even if by easier means, he turns round to look out over the astonishing panoramic view, stretching over the entire Bay of Naples, to the islands and far beyond the city limits. He is, in effect, looking at a three-dimensional map of classical antiquity, a landscape in which he can hear the voices of Pliny, Virgil and many more evocative echoes. The scene is as beautiful, he imagines, as it has ever been. Having drunk it in, he busies himself in drawing the scene as well as noting down what catches his eye. 'The mountaine consists of a double top; the one pointed very sharp, and commonly appearing above any clowds; the other blunt.' He describes the outer, northern wall made by Monte Somma and the ambitious little upstart peak, Mount Vesuvius, now fast overtaking its moribund parent. The explanation for the steep cone was, unknown to Evelyn or his contemporaries, caused by the viscosity of the lava, the result of high silica content. When the consistency of the lava is more liquid, it flows quickly and hence leaves a shallower form or, rarely, an entire lava lake – a landscape existing at spectacularly high temperatures.

Around these peaks, Evelyn wrote, the actual ground is turbulent; 'as we approach'd, we met many large and gaping clefts & chasms, out of which isu'd such sulphurous blasts & Smoake, that we durst not stand long neere them'. The incidence of fumaroles and random volcanic chimneys caught the attention of every traveller, but at a time of recent activity they must have been alarming – who knew when and where they might next burst into life? Arriving at the crater's edge,

I layde my selfe on my belly to looke over & into that most frightfull & terrible Vorago, a stupendious pit of neere three miles in Circuit, and halfe a mile in depth ... The area at the bottom is plaine, like a curiously even'd floore, which seemes to be made by the winds circling the ashes by its eddy blasts: In the middle & center is a rising, or hill shaped like a greate browne loafe, appearing to consist of a sulphurous matter, continually vomiting a foggy exhalation, & ejecting huge stones with an impetuous noise and roaring, like the report of many musquets discharging.

Evelyn and Henshaw stay there, entranced, for hours before taking the relatively easy descent, slithering down through deep ash, which 'at the late eruption was a flowing river of mealted and burning brimstone'. In this breathless, thrilled account, he exaggerated only the measurements.

Their expedition was a reminder, too, both of the rebel Spartacus lurking 'amongst these horid Caverns' within the crater and of Evelyn's hero, 'the learned and inquisitive Pliny', who was to 'adventure his life, to detect the causes, & to loose it in too desperate an approach'. But the most recent event, the eruption of 1631, moves Evelyn to greater eloquence. It had,

> burst out beyond what it had ever don since the memory of any history; spewing out huge stones & fiery pumices in such quantity, as not onely inviron'd the whole mountaine, but totally buried & overwhelm'd divers Townes people & inhabitants, scattering the ashes more than an hundred miles distance & utterly devastating all those goodly vineyards where formerly grew the most incomparable Greco; when bursting through the bowels of the Earth it absorb'd the very Sea, and with its whirling Waters drew in divers gallyes & other Vessells to their destruction, as is faithfully recorded: Some there are who maintain it the very Mouth of hell it selfe.

Evelyn, ever an avid reader, had absorbed his share of eye-witness reports of that terrible catastrophe. Two years after his visit, the city challenged the volcano with its own outburst. For once the reliable opiate of religious ritual and regular popular festivals, often involving theatrical pyrotechnics with fire and elaborate light effects as if mimicking the volcano, failed to keep the urban poor subservient. The civil and political enervation upon which the absentee Spanish Habsburg regime had depended for so long, was suddenly challenged by a Spartacus-like hero of the poor, Masaniello, who led a short-lived revolution in July 1647.

By then, John Evelyn was in Paris and married. His only contact with southern Italy now lay in his notes, from which he composed his account, and in his drawings. In the diaries, not published until the early nineteenth century, Evelyn wrote exceptionally vividly about Vesuvius. All but one of his sketches are lost; the one surviving shows a view from high up over the Bay of Naples and beyond, the scene he had described rhapsodically. Back again in Paris in 1649, having quietly absented himself from republican England, he made a suite of etchings after these drawings, reliving the Italian leg of his Grand Tour. They were published in London but can't have been intended for wide circulation.

Three of the six prints depict the landscape of Vesuvius. One shows the twin peaks, Monte Somma and Mount Vesuvius, rising out of a rough sea of newly set lava while two figures stand gazing at the spectacle as their mules and a guide wait close by. The next, a view down into the depths of the crater, an arena of inhospitable vertical rock faces, focuses on a spouting central feature, possibly the hot magma breaking through the crust. Thomas Henshaw – to whom he dedicated the suite – skulks in a broad-brimmed hat, balanced perilously close to the edge. The third etching is based upon the single surviving drawing. A horseman rides down the mountain towards distant Naples, the city wrapping neatly around the rim of the bay at his feet. A church building to the south perhaps indicates Resina (modern Ercolano, the site of Herculaneum), their starting point. Evelyn has conjured up the scene with scrupulous accuracy but with a hint of poetic sensibility. Kircher may have dissected the volcano, theorizing on the etching plate, but for most travellers the evidence before their eyes was sufficiently amazing in itself.

8. John Evelyn's etching of the crater of Vesuvius. In 1645 Evelyn and his friend Thomas Henshaw found the scene 'engaged our contemplation for some whole houres both for the strangeness of the spectacle, and for the mention which the old histories make of it … one of the most stupendious curiosities in nature.' Henshaw, hardly distinguishable from the rock, gazes fixedly into the cauldron of bubbling lava below. Evelyn's print is disarmingly literal – he was an enthusiastic, but novice, print-maker.

Evelyn's memories of his ascent of Vesuvius remained acute for the rest of his long life, even though he never left England again. *Fumifugium*, written at the time of the Restoration, transforms a smoke-cloaked London into the volcanic suburb to hell. In old age he wrote to remind his friend Thomas Tenison, by then (1692) the Archbishop of Canterbury, of a successful experiment in spontaneous combustion they had both witnessed long ago at Gresham College, then home to the Royal Society. They were still determined to find a link between sulphur and volcanic activity. Evelyn had never forgotten seeing the heaving, fuming acid-yellow deposits that lay on the ground around the Campi Phlegraei, his visit to the sinister, lifeless waters of Lago d'Agnano (a long-dead crater, to be drained in the nineteenth century) and his glimpse of that habitual pantomime ('the old experiment' he called it) at the Grotto del Cane (the cave of the dogs) in which one animal was put in the tiny cave where the poisonous gas at floor level knocked it unconscious. Brought out and plunged into the cold lake water, it immediately recovered. A second dog was left in the cave and soon expired. These diabolical places suggested that the earth was continually venting itself of all manner of vapours and noxious substances. The active volcanic belt across southern Italy and into the nearby islands was a live chemistry set – the difficulty was to know what the laboratory experiment was and where it was leading.

For the early Fellows of the Royal Society – industrious amateurs now home again, such as Evelyn and Henshaw, as well as the stars in a new galaxy of natural philosophers, men like Robert Boyle and Robert Hooke – their close (but rarely firsthand) study of geological phenomena pointed in a number of directions. Considering volcanic activity, Boyle toyed with explanations for the differences between chemical and 'natural' fire. He asked for firsthand reports from those friends and colleagues who had witnessed volcanic eruption in order to illuminate the mystifying but proven fact that 'nature herself doth, by the help of fire, sometimes afford us the like productions that the alchemists' art presents us; as in Itna, Vesuvius, & other burning mountains'. Decade after decade the papers read to the Royal Society and published in its journal *The Philosophical Transactions* provided innumerable, unsorted clues (as well as entirely false trails) for those examining the physical world, many of them readers far beyond English shores. Meanwhile Restoration clerics, pursuing an agenda antithetical to that of the open-minded Fellows, laboured over another theoretical explanation. If the waters of the Flood had really drained away deep into the core of the Earth, the upward thrust of the molten volcanic material evidently signalled an impending apocalypse for the cosmos. It was just a matter of time.

While rational Joseph Addison decried the purported miraculous properties of San Gennaro's blood (which even a man as sober and Anglican as John Evelyn had rather hoped to see 'boiling') he delighted in everything about Vesuvius, including its odder properties – the ash trickling down the cone made him think of sand in an hourglass. In a ringing phrase, John Locke had said that there was 'nothing constantly observable in nature which will not always bring some light with it and lead us further in the knowledge of her ways of working.' With that in mind, Addison's literary acquaintance, the Anglo-Irish philosopher and cleric George (later Bishop) Berkeley, paid two extended visits to southern Italy, on one occasion as a chaplain and the next as a tutor, and forwarded his observations to the learned society in London.

Berkeley, the empiricist, witnessed Vesuvius in action over a period of some two months in 1717. Even though this was not a major eruption in the chronology of the volcano, falling between one in 1707 and another in 1737, it is typical of the lengthy periods in which Vesuvius remained intermittently active (behaving much as Etna has always done) such as later in the eighteenth century and for much of the nineteenth century. On this occasion Berkeley sent his increasingly fascinated reports to the Royal Society, via Dr John Arbuthnot. They quickly appeared in the *Philosophical Transactions* as a truthful record for the Fellows, of an extraordinary event as it had unfolded in front of him. If John Evelyn's account was a breathless paean to an astounding natural phenomenon, Berkeley's was rather more considered, but no less subjective.

George Berkeley's central thesis was that the individual only experiences physical objects through the sensations that they evoke, rather than through their evident reality, their material existence. Thus his picture of the volcano, a dizzying kaleidoscope of effects caught in Berkeley's spontaneous and evocative prose, bore out the central credo of his philosophy, *'esse est percipi'* – to be is to be perceived.

At close quarters, teetering on the very edge of the crater just as Evelyn had done, he saw

> *a vast aperture full of smoke, which hindered the seeing its depth and figure. I heard within that horrid gulf certain odd sounds,*

which seemed to proceed from the belly of the mountain; a sort of murmuring, sighing, throbbing, churning, dashing (as it were) of waves, and between whiles a noise like that of thunder or cannon, which was constantly attended with a clattering like that of tiles falling from the tops of houses on the streets.

After an hour the smoke cleared and offered a view of the crater within which were 'two furnaces almost contiguous'.

Returning some three weeks later, he found everything changed. 'A conical mount had been formed since my last visit, in the middle of the bottom: this mount, I could see, was made of the stones thrown up and fallen back again into the crater.' He saw two gaping holes, one

raged more violently than before, throwing up every three or four minutes with a dreadful bellowing a vast number of red-hot stones … at least 3000 feet higher than my head as I stood upon the brink: but there being little or no wind, they fell back perpendicularly into the crater, increasing the conical hill. The other mouth to the right was lower … I could discern it to be filled with red-hot liquid matter, like that in the furnace of a glasshouse, which raged and wrought as the waves of the sea, causing a short abrupt noise like what may be imagined to proceed from a sea of quicksilver dashing among uneven rocks. This stuff would sometimes spew over and run down the convex side of the conical hill; and appearing at first red-hot it changed colour, and hardened as it cooled, shewing the first rudiments of an eruption, or if I may so say, an eruption in miniature.

He judged himself lucky since the wind direction was favourable, protecting him from smoke and airborne molten lava. 'I had an opportunity to survey this odd scene for above an hour and a half together.'

Back in Naples, Berkeley kept close watch. On June 7 he reported Vesuvius again at full strength; 'it began a hideous bellowing, which continued all that night and the next day till noon, causing the windows, and, as some affirm, the very houses in Naples to shake. From that time it spewed vast quantities of molten stuff to the South, which streamed down the side of the mountain like a great pot boiling over.' Even the northern side was now overflowing. Berkeley was at pains to convey the sheer din that the volcano made, its 'roaring and groaning' audible from the far side of Naples. Imagine sounds, he asked his readers, 'made up of the raging of a tempest, the murmur of a troubled sea, and the roaring of thunder and artillery, confused all together.'

Like Pliny the Elder, his curiosity was so great that with three or four companions he sailed out across the bay to Torre del Greco, to see it from the south. They were astonished by the garish colours reflected in the clouds above the crater: greens and yellows as well as red, lingering on into the darkness when a dull rouging of the sky above the lava flow added to the sinister aspect. Then, having landed in the dead of night, they confronted

a scene the most uncommon and astonishing I ever saw, which grew still more extraordinary as we came nearer the stream. Imagine a vast torrent of liquid fire rolling from the top down the side of the mountain, and with irresistible fury bearing down and consuming vines, olives, fig-trees, houses; in a word, every thing that stood in its way. This mighty flood divided into different channels, according to the inequalities of the mountain: the largest stream seemed half a mile broad at least, and five miles long.

It was exactly the scene that the physicist Giovanni Alfonso Borelli described in his classic account of the 1669 eruption of Etna, with which Berkeley knew his readers were likely to be familiar.

Like almost every volcano aficionado before or since, Berkeley took a risk too far, walking perilously close to the 'river of fire' and having to run for his life, 'the sulphurous steam having surprised me, and almost taken away my breath'. They came down the mountain at three in the morning, but the 'murmur and groaning' went on. Vesuvius was 'throwing up huge spouts of fire and burning stones, which falling down again, resembled the stars in our rockets.' The flames, gushing out of the crater, shot up in single, double and even triple columns. The next morning brought 'a sort of eclipse. Horrid bellowings this and the foregoing day were heard at Naples'. One day the mountain was invisible under its black cloud, the following day their house in Naples was clothed in ash.

Astonishingly after two or three more days, Vesuvius was utterly calm and neither smoke nor flame could be seen. Tantalisingly, he told his correspondent,

> *It is not worth while to trouble you with the conjectures I have formed concerning the cause of these phænomena, from which I observed in the Lacus Amsancti, the Solfatara, &c. as well as in Mount Vesuvius. One thing I may venture to say, that I saw the fluid matter rise out of the centre of the bottom of the crater, out of the very middle of the mountain, contrary to what Borellus imagines ... I have not seen the crater since the eruption, but*

design to visit it again before I leave Naples. I doubt there is nothing in this worth shewing to the Society; as to that, you will use your discretion.

Facts were piling up, like heaps of ejected volcanic fragments, sensations too, but the conclusions remained elusive.

George Berkeley's vivid firsthand observations, offered to the wide and international readership of the learned society's journal, may not have provided new geological insights but are believed to have inspired James Thomson's poem 'Summer' thus;

> The infuriate hill that shoots the pillared flame:
> And, roused within the subterranean world,
> The expanding earthquake, that resistless shakes
> Aspiring cities from their solid base
> And buried mountains in their flaming gulf.

Thomson's poem was published in 1727 when he was living in London. For him, and all his readers, the future Bishop had conjured up a maelstrom of sensual effects alongside his careful, factual account, and offered graphic proof of the psychological and aesthetic power of that turbulent, literally awful landscape.

3

WILLIAM HAMILTON – MADE BY VESUVIUS

With the arrival of the first monarch to reside in Naples for two centuries, following a long succession of hated Habsburg viceroys, a semblance of confidence returned to the demoralised city. 1734 saw the installation of a new Spanish ruling dynasty, the Bourbons, and in 1738 King Charles VII began to build himself a relatively modest palace at Portici, bridging the road that ran south from the city, between Vesuvius and the sea. Where the king led, a suburban building boom followed. The architects of the one hundred and twenty or so villas to be built around the Golden Mile, as it became known, played upon the idyllic scenery just as the Romans had done. These houses embraced the landscape, often facing two ways, their terraces and balconies offering picture-perfect views of both mountain and Mediterranean. Some went so far as to catch Vesuvius or, in the opposite direction, the Bay of Naples, within decorative external screens. The effect was to frame the vista, rather like a showcase in which the volcano was the precious object on display. And these houses were built of the materials to hand, volcanic materials; constructed of tufa rendered with pozzolano, the latter an ingenious way of improving the poor quality stone. Paving was of *piperno*, tough basalt.

This most elegant and hedonistic suburb gained a certain indefinable frisson from the continuous risk (and thrill) of volcanic eruption playing against the owners' implicit self-assurance – in their own power, high birth and status, wealth and ingenuity. Doubtless they had well-prepared escape plans should the need arise. Yet a ringside seat was not without its dangers for the population, as the king implicitly acknowledged by paying an immediate visit to the heavily decorated Baroque chapel of San Gennaro in the Duomo, at the heart of the dense medieval city.

Three years later came a major eruption that, fortunately, spared Naples but gave aspiring residents of the Golden Mile a taste of the turbulent locality. As if to calm their nerves, a measured account was published. The author, Francesco Serao, admonished his readers; 'Let us now keep close to the Truth'. He briskly dismissed arrant nonsense like the suggestion that Vesuvius had erupted in disapproval of the arrival of the Bourbons in person. Soon after that Serao published *Della Tarantula*, a measured challenge (made with the blessing of the Church) to the ancient and pagan rituals nominally associated with the deadly spider. Once, it had led to festive saturnalia but, by this time, the exorcism was taking musical form, the eponymous tarantella, a dance resembling a frenzied eruption – of the human body. Faced with Vesuvius, Serao's analysis was equally and determinedly scientific (although he also kept one eye, and his prose, well fixed upon continuing royal patronage – for he was the king's doctor). He believed that the event was triggered by chemical reaction, a 'close Mixture' of minerals proving highly flammable since, as he put it, Neapolitan soil is 'full of the Seeds, or first Principles of Fire'.

Serao was the first to use the word lava, derived from the Latin *labes* for fall or slide. In an attempt to add a positive note, his account ended with some safety measures, suggesting that people build dykes and ditches or divert and divide the main flow of molten material – as had proved effective at Etna. As *Natural History of Mount Vesuvius* it was translated into English in 1743, with a plate showing the volcano in section. Soon after, Thomas Nugent's 1749 guide to Italy described the charms of Naples to the British, for whom 'there cannot in all respects be a more agreeable place to live', conveniently ignoring the fact that famine and natural disaster were endemic though, in passing, cautioning his readers about potential disturbance from 'the eruptions of the neighbouring mount Vesuvius, together with the earthquakes'. Like the insouciant Neapolitan nobility, pleasure-seeking Grand Tourists came to look on volcanic activity as a bonus rather than a hazard. They certainly did not view Vesuvius as an Enlightenment project, as Serao had and many others were to.

For William Hamilton, the new Ambassador to the Kingdom of the Two Sicilies (consisting of Sicily and the entire southern part of Italy up to the Papal States), Vesuvius was the making of the man; and he was the man who made Vesuvius. It was largely his doing that the volcano came to fascinate late-eighteenth-century Europe to such an extent. A man of incisive intellect and considerable self-knowledge, hardly the caricatured cuckold he would become, Hamilton and his first wife Catherine arrived in Naples in November 1764. Yet there is no clear explanation as to what in particular triggered Hamilton's obsession.

A younger son of an aristocratic Scottish family, he spent some years overseas as a guardsman until marriage, a seat in

9. The Villa Campolieto at Ercolano in the late 1950s. The villa was designed and built from 1763 onwards by Luigi Vanvitelli and, like its neighbours on the Golden Mile, framed the glorious landscape with arcades and screens, passages and terraces.

parliament and his appointment as an equerry to George III brought him home. He showed signs of being an autodidact, learning the violin and, advised by Horace Walpole, busily buying and selling paintings. But his posting to Naples as ambassador was ostensibly for the sake of Catherine's health which it was supposed would benefit from a warm, dry climate. There, Hamilton found himself at the court of thirteen-year-old King Ferdinand IV, on the throne since 1759, and whose puppet master was the ageing Bernardo Tanucci, of whom Leopold Mozart, Wolfgang's father, was to note, 'this Prime Minister is really a King'.

The Hamiltons arrived to find Naples in the double grip of famine and disease. (Shockingly, the city experienced a cholera epidemic as recently as 1973.) The regime had been lurching back towards malign autocracy since Charles had departed to assume the throne of Spain, installing his then eight-year-old son, Ferdinand, as king of Naples. It was as if, playing to those primitive notions of divine punishment that Serao had so firmly scotched, Vesuvius' renewed activity was attuned to the current Neapolitan mood. The court of the now-teenaged monarch often escaped north to the gargantuan new royal palace at Caserta for prolonged and profligate hunting expeditions. Hamilton was often required to join the party. Outside the vacant royal palace, an ornamental pyramid of food – the *cuccagna* – was assembled. The Neapolitan poor, the *lazzaroni*, were then invited to rip it apart. It was part festivity, part cruel spectacle. Their ravening hunger was evidence that despite the astonishing fecundity of the surrounding countryside, the social order was entirely feudal. Naples was the most densely populated capital in Europe and the most medieval in both fabric and attitude. An apt reflection of the febrile political situation below, Mount Vesusius reared up over the ancient Greek settlements of Neapolis (new town), its predecessor Parthenope and their outlying regions, an unpredictable dominatrix.

At the foot of the volcano were the recently rediscovered towns of the dead: Herculaneum and Pompeii. The latter had been discovered in the 1590s, the former in 1709. On Charles VII's arrival in Naples, subterranean work at Herculaneum recommenced and from 1748, excavations began at Pompeii. All spoils were regarded as endowments for the Bourbon–Farnese throne rather than as the cultural and intellectual property of a wider Europe. There was no consideration of their archaeological importance or the importance of where they had been found, for these were objets d'art garnered for the glory of the regime to be added to the Farnese family collection which was now brought from Rome and elsewhere to be housed in Naples, much of it at the royal museum-cum-palace at Portici.

Fortunate visitors who gained privileged access to the tunnels below Herculaneum or to the growing collections in the palace (from 1758 known as the Museo ercolanese) were forbidden to linger long enough to note or sketch what they saw. No outsider could study these treasures in detail, far less publish them.

For Hamilton, the stage was set. He began buying Greek vases, at the time erroneously called Etruscan, from the moment he arrived in Naples. Within six months he had a notable collection and was already considering publication. The antiquities of Pompeii and Herculaneum awaited but Vesuvius was visibly and audibly stirring. The telltale signs, noise and billowing smoke, began in June 1765 and he was instantly and irrevocably smitten. The volcano was, he wrote, 'in labour' and he began to watch its pains like an anxious first-time father.

Hamilton first climbed Vesuvius in the snow. By November 1765 the mountain was furiously spewing stones (including a vast 'bomb' which only just missed him) but the eagerly awaited lava flow did not materialise. Many Grand Tourists witnessed such early warning signs and went home, convinced that they had seen an eruption. But on Good Friday, 28 March 1766, Hamilton witnessed a fully-fledged eruption – he had caught Vesuvius at the optimum moment and revelled in his good fortune. On 12 April he spent an entire day and night on the mountain, alone but for his guide. On another occasion he was acting cicerone to a young German prince, Leopold III Friedrich Franz of Anhalt-Dessau, with his architect and others. Vesuvius would seize their imaginations almost as forcibly as it had Hamilton's, with remarkable results.

By June, Hamilton felt confident enough to report back to the Royal Society upon 'the many extraordinary appearances that have come under my own inspection'. It was the first of numerous communiqués from Vesuvius (and, on one occasion, Etna) sent to London over the next thirty years. He was duly elected a Fellow of the Royal Society within the year.

The Hamiltons' official residence in Naples was in a former convent on the seafront, the Palazzo Sessa. (The house survived to become the centre for the Jewish community. It has housed a synagogue since 1864). The couple had no children but the rooms quickly filled with possessions; it was always both their home and a museum. Later on, as Catherine's frail lungs weakened, her husband designed her an upper-level balcony room, its bow windows opening onto a panorama of the Bay of Naples. Every surface was mirrored, including the backs of the doors, so that she could believe herself to be outside. After her death, Giovanni Battista Lusieri painted the view that she had so enjoyed – the translucent, still water contrasting with the intensely animated foreground. From that angle Vesuvius could not be seen, but then she had never shared her husband's obsession and never climbed the volcano, even in her stronger years.

In the first years of his posting to Naples, Hamilton had, in effect, given himself the role of cultural emissary to replace the distasteful one of political representative to a rotten regime. He inscribed a Latin motto *Ubi bene, ibi patria*, 'where I am at ease, there is my native country', onto a wall, for the sober, endlessly enquiring Sir William had consciously stepped into the shoes of Pliny the Elder, with his insatiable hunger for knowledge and an attentive nephew. It wasn't long before London wits began to call Hamilton and Charles Greville (his own nephew) 'the two Plinys'.

Hamilton also took a lease on the modest Villa Angelica at Portici. Conveniently close to the royal museum and palace (and woods), situated between Herculaneum and Pompeii and offering an unimpeded view of Vesuvius, Hamilton had an ideal observatory. As we know from his artist visitors' watercolours, outside the windows lay a large garden, a naturalised riot of myrtle, broom and vines shaded by cypresses and pines and, beyond that, rising towards the foothills of the mountain, plantations of evergreen oak. All growth was manic, fed by the rich volcanic ash. Hamilton tended his jungle himself, 'cutting Walks & making a kind of labyrinth in a Shrubbery' as Catherine wrote to her cousin, William Beckford.

10. The Hamiltons are portrayed sitting companionably in their 'casino' at
Posillipo in 1770. East across the Bay of Naples, Vesuvius puffs peaceably.
Catherine, an admired musician, plays her pianoforte and William listens, his
many interests conveyed by books, paintings, a 'basalt' vase, an antique bust of
Serapis and his violin. The official looking papers are communiqués but also,
perhaps, include his letters to the Royal Society with current observations on
the volcano that was to be his lifetime study.

Vesuvius was the perpetual backdrop, shifting its tones with the weather and the season through lavender to slate to violet and back to the purest harebell blue, then as now. Seen at a distance it never appeared as it really was, forest green shading to dun brown. From the windows, balconies and, perhaps, even the roof of his *villa maritima*, Hamilton kept vigil over the volcanic cone behind the house. He watched its every shift of mood, both with the naked eye and with the latest and finest telescopic lenses.

Closer to the city, but with direct access to the sea and offering clear views back to Vesuvius from the west, the Hamiltons' Casino overlooked the beach at Posillipo. A refuge from the summer heat of the city, it had no more than three rooms and a kitchen, with a cool terrace beyond. David Allan painted them there in 1770. Catherine, a notable musician, is at her keyboard and William sits beside her with an antique bust behind him; to each, an attribute. Outside the open window, Vesuvius smokes pacifically, perhaps the artist's hint that Hamilton was already collating his *Observations on Mount Vesuvius, Mount Etna and other volcanoes*. Eventually the little house became the Villa Emma. It survives, its identity unsuspected until quite recently, the terrace still in evidence.

Hamilton's very first letter to the Royal Society in 1766 pondered the link between the weather and volcanic action, wondering if rough seas might force their way into 'crevices' and thus into the core of the mountain, causing 'fermentation'. He was repeating the antiquated notion that the waters of the Flood had seeped into the core of the Earth. Hamilton also described the cottages and even the vineyards on the volcano's side all bedecked with talismanic images of San Gennaro. But from then on Hamilton reported the facts, with few digressions, personal opinions or picturesque asides (though he occasionally added drawings). He wrote in the spirit of the society's motto, offering information to those 'more learned in Natural Philosophy' than he. He did not embellish or postulate; he merely described what he saw. Unlike many of the essentially collaborative publications that poured out from the Fellows of the Royal Society from the 1660s onwards, Hamilton was not editing or collating the reports of others, since everything he wrote about he had witnessed himself. Obligingly, Vesuvius responded to his rapt attention; eruptions followed in rapid succession.

In October 1767, Hamilton had been almost at the edge of the crater when, hearing a violent noise, he looked up and the mountain seemed to split apart in front of his eyes. From the rent 'a fountain of liquid fire' shot up and then rolled in his direction. He and his trusted one-eyed guide Bartolomeo Pumo (known as the Cyclops of Vesuvius), did not stop running for three miles. Arriving in Portici, he wrote, 'I found my family in a great alarm', the villa shaken to its foundations, its doors and windows swinging open. Hamilton's mother-in-law Ann Barlow quickly wrote hoping 'to hear that all your fine & horrible sights are over, & that the Mountain is all quiet again.' She also hoped he would be more careful in future.

In the coming days, Hamilton returned to the crater over and over again with different companions. One was the Rome-based antiquarian and art historian, J. J. Winckelmann, the greatest authority on the art of the classical world, who was murdered in Trieste soon after. Another was Frederick Hervey, the future Bishop of Derry and Earl of Bristol, an enquiring geologist but a dishonest patron who conveniently forgot his promises to many artists. Hervey was severely scalded by flying detritus from Vesuvius.

Word of the 1767 eruption soon reached North America. John Morgan from Philadelphia, a founding father of American medicine who had visited Vesuvius in 1764, heard about it from an 'English Gentleman Residing at Naples' and published the account in the *Transactions of the American Philosophical Society* soon after. Morgan's correspondent wrote how his household had retreated in terror to the rooms at the rear, built into the hillside, so great were the shock waves. But once the lava began to flow, 'from a seeming opening of the whole side of the mountain at once', the incessant shaking stopped, although it still seemed wiser to stay indoors since 'a tumultuous concourse of people' had taken to the city streets with images of the Madonna and saints. The king and court had fled and were reportedly heading straight to safety at Caserta. A week later Morgan's unnamed informant rode out to take a closer view but 'was obliged to gallop home with my eyes shut, as I could no longer open them from the pain these ashes put me to.' Every account added another telling detail to the picture.

Obliged, like Pliny, to work in a crude modern world of power struggles and chicanery, Hamilton took comfort in his congenial tasks. Periodically the rooms of the Palazzo Sessa must have emptied as the collection was crated up for sale or presentation to the British Museum or the Royal Society. The first shipment arrived safely in 1772 and the museum spent £8,400 on the majority of the Greek and Roman antiquities on board. Before they left his charge, Hamilton arranged publication of his finest Greek (Etruscan) vases in several volumes, for which he enlisted the services of the dubious self-styled Baron d'Harcarville, as well as the erudite Winckelmann. Soon Hamilton began to report back on his visits to archaeological sites, in 1774 describing new finds at Pompeii to the Society of Antiquaries. Such promotion would alert potential purchasers, for his resources were draining away fast, due both to his own extravagance and to the Treasury's failure to reimburse him for his official expenses. His starting salary as envoy was £5 per day, which rose to £8 when he became a grander figure, Minister Plenipotentiary, in 1767. By the mid 1770s he was heavily in debt due to the lack of payment from the Civil List.

The geological specimens came in four batches between 1768 and 1779 but they fell between categories – neither particular enough for a cabinet of curiosities nor aesthetically suitable for the new museum. Sir Hans Sloane gave 10,000 rocks and minerals to the British Museum but the emphasis was on their aesthetics, especially when nature spectacularly imitated art. Mineralogy had less standing in Britain than in other northern European countries and Hamilton's chests of rocks and minerals were never analysed, his measurements neither verified nor collated. Their most public, and enduring, display was to be in Pietro Fabris' plates to the *Campi Phlegraei*. There, the most eye-catching items were shown beautifully arranged in trompe l'oeil display. There was, perhaps, a dash of irony intended for, in his eyes, these were trophies equal to the much-celebrated antique terracotta vases.

The formal divide between the men of science and the assiduous amateur was to widen in the early nineteenth century but for now, William Hamilton straddled it comfortably. In 1770 he received the highest honour of the Royal

Society, the Copley Gold Medal, awarded for his account of Mount Etna. Two years later he was knighted for his official duties. Meanwhile a copy of Hamilton's 1772 *Observations* had reached Voltaire, who responded by quoting Pliny at him and comparing volcanic activity to the work of ants.

The Villa Angelica became a shifting salon for an international array of artists, writers, connoisseurs and thinkers living in or passing through Naples. In addition, the Hamiltons, as they found to their cost, were magnets for every Grand Tourist in town, be they dolt or genius. In the early days Catherine held weekly salon chamber concerts, at which she played harpsichord, and her husband (and members of their staff) violins and cello. At the Villa Angelica in 1770, the fourteen-year-old Mozart performed for his hosts (whom he had already met in London in 1764) against a percussion accompaniment of volcanic activity. Leopold and Wolfgang visited Vesuvius themselves and, twenty-one years later, Mozart's final opera *La Clemenza di Tito* began with the Emperor Titus moved to divert the gold, donated for a temple in his honour, to the victims of the volcano. Musician Dr Charles Burney described a soirée interspersed with 'the sight and observations of Mount Vesuvius, then very busy' but Lady Hamilton was perfectly used to playing against the volcano, which 'now & then treats me with an explosion'.

On Burney's visit Hamilton lent his guests high-precision lenses and telescopes to focus upon the activity two miles away in the crater. He was determined to share his passion for Vesuvius with others, even those who would never reach Naples. So he designed a kind of domestic *son et lumière* – a shorthand version of Vesuvius incandescent – involving streams of lava, outbursts from the crater and thunderous sound effects, all generated by a painted transparency in a box, backlit through a perforated clockwork-driven cylinder housing a lamp. It was an early moving picture. In 1768 he sent the apparatus with instructions to Dr Maty, who held positions at both the British Museum and the Royal Society. Soon London audiences were sharing the 'awfull phenomenon'; one participant considered 'nothing wanting in us distant spectators but the fright that everybody must have been fired with who was so near'. Hamilton sent another such 'mechanical contrivance' to David Garrick. When in London Hamilton obtained an electrical device that simulated lightning, the sort of experiment that Benjamin Franklin and his Birmingham friends had greatly enjoyed. It was still novel in Italy, however, and much admired by the Neapolitans.

But some felt Hamilton's volcanic obsession had gone too far; in 1773 Horace Walpole wrote affectionately to him, 'Ransack Herculaneum, sift Pompeii, give us charming vases, bring us Correggios and all Etruria, but do not dive into the caverns of Aetna and Vesuvius … it is better to be an antiquary of taste, than a salamander that had passed a thousand ordeals – I am sure Lady Hamilton is on my side …'

Hamilton's geological expertise was now widely recognised and his letters to the Royal Society translated into French and German, Russian, Italian, Danish and Dutch. Savants, geologists-to-be, sought him, men such as Horace-Benedict de Saussure, a young Swiss Alpine expert who in the 1780s was the first to measure and scientifically record the characteristics of high altitude and mountain geology. Hamilton guided him up Vesuvius during his stay in Naples in 1772–73 and de Saussure offered a reciprocal sublime experience, a tour of Chamonix in July 1776. Hamilton was alert to the importance of marine fossil evidence in the strata and now tended towards the Vulcanist view, believing the earth to be the product of intense heat as opposed to the work of the oceans, as did Neptunists. Nevertheless, while 'aware of the danger of systems, I have kept clear of them, and have confined myself to the simple narrative of what I have remarked myself.' He collected information on the volcano Hekla, in Iceland, and on the Giant's Causeway and its army of basalt columns. French 'naturalists' ('my brethren *Volcanistes*') shared their theories with him though, he commented acidly, they might consider visiting 'our active Volcano' to be shown what they did not know. Meanwhile, 'I shall content my self with collecting facts & let who will form them into a System afterwards.' Such geological arguments were best left to others, even if he did have a view.

But it was Hamilton's commission to Pietro (or Peter) Fabris to add coloured plates to his printed letters that brought him the greatest renown as a student of the volcano. The *Campi Phlegraei* of 1776 is one of the most beautiful books ever published. It had a bilingual text, in recognition of French pre-eminence in science. Hamilton's letters to the Fellows of the Royal Society, illustrated by Fabris's hand-coloured engravings, showed every observable characteristic of Vesuvius and the wider volcanic landscape, in winter and summer. In several plates Hamilton himself appears, nattily dressed, to underline the highly personal nature of the enterprise. He had ploughed £1,300 into the project (almost half of his annual salary, representing well over £80,000 in modern value) and so a donor portrait here and there was hardly self-indulgent. After the 1779 eruption, he added a supplement with five plates. The book, necessarily only a limited edition, was Hamilton's testament to the fabulous phenomenon that was Vesuvius. In the early stages of the eruption of 1778 he confessed to his addiction; he could not miss a moment, for 'tho' I have seen this Phenomenon so often I cou'd scarcely leave that Curious Spot before day break'.

More conventionally, Hamilton commissioned artists, and encouraged others to do so, to capture Vesuvius in every mood, season and hour. They rose to the challenge, revelling in the temptation to exaggerate the effects, ratcheting up the drama to outdo each other on canvas. In these versions, generally night views under a full, silvered moon, Vesuvius looked like a terribly wounded creature, vermilion lacerations aflame. Onlookers were pictured in the glow, unfeasibly close to the inferno. As well as a number of compatriots, Hamilton patronised the Frenchmen Pierre Jacques Volaire and Antoine Ignace Vernet, the German Jacob Philipp Hackert (who would take Goethe as a pupil) and the Italians Lusieri and Pietro Antoniani. After Fabris died in 1792, Hamilton commissioned Xavier della Gatta, a Spaniard, to paint the 1794 eruption. One particularly handsome landscape, from a location far across the bay, catches the immensity of the towering ash cloud and here and there the random flashes known as *ferilli*, what Hamilton called 'zig-zag lightning' – a particular feature of violent eruptions, as he told the Royal Society. Like the British watercolourists, della Gatta brought a calm and measured eye to bear on the scene. His understatement conveyed even greater drama: oddly, his painting comes closest to the moving images of the eruption in 1944.

Some headed for Naples in search of extreme emotions, rather than aesthetic impressions or factual observations. Reaching a deserted Naples in a torrential November

11. Pietro Fabris depicts the eruption of August 1779. Rather than show Grand Tourists in the foreground, he has added Neapolitan folk, two popular themes in one image.

rainstorm, William Beckford, William Hamilton's cousin, found Vesuvius blackened. No golden dusk or moonlight, let alone tarantellas or eruptions met him. But the next morning he stepped onto his balcony to see the volcano 'rising distinct into the blue aether' above a 'world of gardens and casinos'. His view was calm and lovely with nothing of the Sublime about it but Beckford soon found the volcano recurring in his dreams. It was perhaps the earliest recorded appearance of Vesuvius in the subconscious (Freud played with the notion later on).

Every artist who hurried south to see the erupting volcano took subtly different readings of what they saw. Some depended upon their own imagination; while J. W. M. Turner, prevented from travelling during the years of the Napoleonic wars, used sketches and prints by J. R. Cozens, the artist who was accompanying William Beckford, and so distanced himself from the scene by several removes. It was many years before he got there.

On his own arrival in 1774, Joseph Wright of Derby regretted that he did not have his close friend, the Derbyshire clockmaker, Lunar Society member and geologist John Whitehurst with him. Had he been on Mount Vesuvius, Wright was sure that 'his thoughts would have center'd in the bowels of the mountain, mine skimmed over the surface only'. And at this moment the volcano was showing promising signs of activity, ''Tis the most wonderful sight in nature.' But Whitehurst did not travel further than London (having been appointed Keeper of Stamps and Weights) and his long promised *Inquiry into the original state and formation of the earth* (1778) depended on observations made around the Derbyshire Peak District, where within a tiny compass was a fantastic cornucopia of stratified rocks, minerals and fossils, hidden underground in the mines and dramatic caves that threaded the area. Based upon all this, he wrote one of the key texts of the Enlightenment. But the inescapable conclusions left the religiously orthodox Whitehurst struggling to set his 'Subterraneous Geography' alongside the events in the Book of Genesis. He separated them, writing an Argument, which tended to support his conservative, literal theology, and an Appendix in which he laid out the evidence of nature. Josiah Wedgwood was one of many who was shocked by White-hurst's tenacious efforts to preserve 'the mosaic account beyond all rhime or reason.'

Wright visited Vesuvius continuously over the years, but only in his imagination. As Erasmus Darwin, another member of the Lunar Society, that fertile intellectual circle of Midland scientists, manufacturers and thinkers whose meetings were held at the full moon (for ease of travel) wrote, 'Wright's bold pencil from Vesuvius' height,/ Hurls his red lavas to the troubled night.' Darwin, Charles's grandfather, was Wright's doctor, and treated him for depression. Wright's view of Etna, which he had never visited, showed it as a distant, pale shape over the rooftops of Catania, intended perhaps as a quiet companion piece to a flaming view of Vesuvius, all scarlet and black. At times Etna seemed to be the useful counterpoint to Vesuvius, oddly less thrilling because it was so dependably and regularly active, a kind of sober sibling to the excitable mainland volcano. Wright painted some thirty versions of Vesuvius, showing it erupting with increasing violence; Darwin must have viewed the furious energy of the series as symptomatic of his patient's fluctuating mood, rather than merely as impressive works on canvas. In 1985, Andy Warhol painted a huge and primary coloured Vesuvius, equally unconstrained by reality. He too published his version in an edition of multiples.

Hamilton's enthusiasm and practical help extended to Joseph Wright's travelling companion, the draughtsman John Downman, who saw gloom in Vesuvius, painting a chill, monochrome view of the rubble-strewn Atrio dell'Cavallo; and to John 'Warwick' Smith whose fresh watercolours became the prints illustrating *Select Views in Italy*. Thomas Jones caught unconventional aspects of Naples, hidden corners, as well as composing entirely conventional landscapes (including a large one for Hamilton). During the 1778 eruption he was shocked to see the demise of the famous vineyards and noted in his diary how 'bunches of grapes shrivel up like raisons [sic] and the leaves wither & take fire upon the Approach of this tremendous burning River'. He judged the molten lava to be 'about as wide as the Tiber at Rome'. In 1783, by which time Jones been in Italy for six years, Hamilton offered him the billiard room at the Palazzo Sessa as studio space. He had turned down Jones's offer to be his draughtsman in earthquake-hit Calabria since he had but twenty days 'to come at the truth', he told the Royal Society. But he left Jones the *Campi Phlegraei* to leaf through in his absence. Such acts of generosity seem to have come easily to Hamilton once he sensed a fellow enthusiast.

Nothing in British geology resembled Vesuvius, but the industrial processes that were making landscape into theatre at home seemed to offer a new way of looking at the volcano, and vice versa. Views of smelting furnaces and burning forges in Coalbrookdale and Bedlam were comparable to the volcanic ejection of smoke and flames – always seen at their best by night. On his return from Naples, Wright of Derby showed moonlight washing the clouds above the cotton mill at Cromford, its hundred windows aglow with warm oil light, just as he (and others) would show a clear cold moon above the Bay of Naples, as a descant note to flaming lava streams. Smoke spiralling from the spoil heaps around abandoned coalmines in the West Midlands reminded travellers of the sweating sulphurous landscapes around Pozzuoli. Yet aesthetic taste had not always embraced industrial processes; in the late seventeenth century the dense white smoke produced while firing salt-glazed ceramic wares was judged 'extremely disagreeable, not unlike the smoke of Etna or Vesuvius'.

Sir William Hamilton, who had charge of his wife's extensive estates in Wales, near Milford Haven, and spent time there when on leave, was familiar with industry and the metaphors came effortlessly to him when he reported on Vesuvius to the Royal Society. The first lava flow he witnessed had brought glass-making to mind while the final eruption, in 1794, produced a thunderous roar like 'that noise which is produced by the action of the enormous bellows on the furnace of the Carron iron foundry in Scotland' – he had been there too.

Joseph Wright portrayed John Whitehurst, sitting inside bent over his book, with Vesuvius smoking heartily outside the window. But both painter and subject were in Derbyshire. Visitors to Naples often took home portraits of themselves (and usually Vesuvius) as well as paintings showing the Bay of Naples, sites of classical antiquity, folkloric scenes (showing

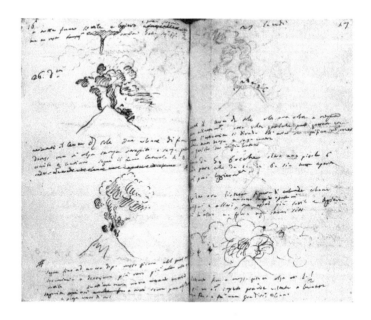

12. These lively drawings and text in pen and ink come from the pages of Antonio (Padre) Piaggio's diaries of Vesuvius. The elderly former papal librarian kept a daily record at the behest of Sir William Hamilton, who for many years paid him to keep an eye on the volcano. Vesuvius was rewardingly active on this occasion – it was 1779, the same eruption that Fabris depicted – and Piaggio's hand can hardly keep up with the speed of the activity and rapid changes he is witnessing, happily from a safe distance.

traditional costume, music and dance, especially the tarantella) and views of Vesuvius erupting by day and night (paired images were popular). An army of Neapolitan commercial artists supplied a growing market. The Scottish painter David Allan showed one in his studio, industriously painting to order. Any Georgian country house from which a Grand Tourist had embarked would now have a parade of gouaches marching up the stairs or clustered like a stamp collection in the drawing room. They were to be viewed sequentially, like flipping the pages of an album. These workman-painters did not fuss about chronology or accuracy; each eruptive view had a date, entirely arbitrary but reassuring to the purchaser, who liked the fallacious authenticity. The market was as insatiable and undiscerning as the dull but well-born travellers with whom Hamilton was frequently forced to toil up Vesuvius.

Hamilton knew his limitations. He aimed to inform those with more expert knowledge, whether of chemistry or geology. As his reputation grew, his findings reached a widening public through periodicals such as the *Gentleman's Magazine* and the *Edinburgh Review*. To help him keep in touch, since his duties as Envoy Extraordinary and Minister Plenipotentiary to the (Spanish ruled) Kingdom of the Two Sicilies made increasingly heavy inroads on his time now that Spain and England were at war in the Americas, from 1779 he paid an annual £20 stipend to a former Vatican librarian to record Vesuvius, day by day. The journals kept by Father Antonio Piaggio until 1794 filled eight volumes which Hamilton presented to the Royal Society where they remain, 'a complet Diary or Journal of the operations of Vesuvius'. They are a touching record, meticulously following his patron's instructions to make 'observations on the smoke of Mount

Vesuvius … and make daily drawings of it'. These were the years in which, at home, Hamilton would become the subject of derision as the dupe of a scheming woman before being caricatured as a pathetic cuckolded husband. Yet, consistently, Hamilton was steadfastly pursuing his mission to report on the minutae as well as the wider picture, of Vesuvius and its works.

Sir William was delighted with Piaggio's diaries since 'altho' they contain some strange ideas, as he is a little of the Rosicrucian [they] are wonderfully curious.' While the volcano was inactive Hamilton asked Padre Piaggio to make drawings from Herculaneum for a publication on the archaeological site. Crucially, this was to include material deciphered from the badly damaged scrolls from the library in the Villa of the Papyri. At the Vatican Padre Piaggio had invented a device to unroll them at snail's pace, without incurring damage. But he died in his eighties, a defeated old man. There would never be a book, for the Dilettanti Society almost immediately lost all the preparatory material.

As his own confidence grew, Hamilton longed for a scientifically literate colleague with whom to share Vesuvius. Joseph Banks, President of the Royal Society from 1778, though primarily a botanist, became Hamilton's sounding board – but he never found time to visit Naples. Hamilton told him of his growing conviction that 'Volcanick operations are much more powerfull Agents than hav[e] been hitherto allow'd' and that many South Sea islands were the result of volcanic activity. (Several in the Pacific came to bear Banks' name.) Hamilton confided his growing frustration to Banks, considering most vulcanologists to be lazy minded in the face of clear evidence; his observations had led him to change his views from Neptunist to Vulcanist – why did they not do likewise? One Italian expert had realised his mistake, 'but thinks it too late to take up a fresh System particularly as he has many of his books to dispose of still'. Thus were mistakes compounded and facts distorted. Yet the 'year without a summer', 1783, was rightly ascribed to the effects of a prolonged series of fissure eruptions at Laki in Southern Iceland (in the same geological region as the volcano which disrupted the world, albeit briefly, in 2010), which then spread their malaise – physical, economic and finally political – across Europe.

Vesuvius had helped to distract Hamilton from his wife's deteriorating health and the growing British tensions with the French in the West Indies. By mid September 1778 another major eruption was signalled. From his windows at Portici, 'not two miles in a strait line from the Crater', he focused his Ramsden telescope on the action, transfixed for almost three hours. But even with the best lenses, he was too far away. Around midnight, he set off to see it at close quarters and spent all night on the mountain. How sad that he was not there with his friends – Joseph Banks and his charming Swedish librarian Daniel Solander, and his nephew Charles Greville, a discerning collector of minerals from around the world – instead of those who 'have in general so much fear & so little Curiosity that I had rather be alone.'

By 1794 he would have ascended to the crater almost seventy times, more often than not in tedious company, in his long alternative career as *cicerone* to Vesuvius. Typical, if unusually entertaining, was a Neapolitan duke who fled in abject terror as the red hot lava set light to juniper bushes, assuming that Vesuvius had erupted. As he ran he screamed to an engineer in his party to take care of his 'beloved duchess'. The

13. This seventeenth-century painting of San Gennaro shows the saint in his bishop's robes and (anachronistically) holding a phial of his own blood. If the substance miraculously turns to liquid, the citizens of Naples can sleep soundly for another few months, safe from the eruptions of Vesuvius, as from plague or famine. Recently, an online version of this painting has been doing the rounds –the face now that of Berlusconi.

higher born the individual, often the worse their behaviour. The Habsburg Emperor Joseph (the king's brother-in-law) had no compunction in beating Hamilton's trusted guide, Bartolomeo, with his cane. The future Tsar Paul I of Russia claimed his lungs were too weak to climb up to the crater, while the Duchess was too fat and her shoes not strong enough (Hamilton had, luckily, suggested she take spares). But at least they enjoyed themselves. Even Ferdinand and his Queen, Maria Carolina (Marie Antoinette's sister), made an effort to view the eruption of 1771 from below. Fabris depicted them with poetic licence, standing with their backs to the artist, illumined in the glow of the flames.

Banks kept Hamilton's spirits up, pointing out that his pen was in tune with Vesuvius 'when the one Belches out Fire & smoak … the Mountain Evacuates Learning & [in]formation'. Banks reminded him that his 'Philosophic researches' were the ideal way to 'banish Enui'. The latest eruption even distracted from the heightened political tensions of the summer of 1779 as the French took Grenada from the British during the American War of Independence. As Horace Walpole wrote to Horace Mann; 'What are kings and their pop-guns to that wrath of nature!' but he wondered if revolution might satiate Hamilton's appetite for volcanoes: 'Are there not calamities enough in store for us, but must destruction be our amusement and pursuit?'

In Naples a statue of San Gennaro now stood guard on the Maddalena Bridge. When Vesuvius erupted in September 1778 Thomas Jones passed two porters carrying the saint's bust beyond the city limits, followed by a great crowd, 'in order to put a stop to the farther depredations of the Lava', as he put it. Strictly speaking, the saint was outside his jurisdiction and the patron saints of other parishes were intended to take responsibility. In 1779, the actual head of the saint came out onto the streets, always a desperate last move. Vesuvius was a pressure gauge for geological force but also for the desperation of the Neapolitan poor. Hamilton began to wonder if Naples did not face more trouble 'from the irregularities of its lower class of inhabitants than from the angry volcano'. His servant Gaetano faced all the eventualities by combining tattoos of I.C. (Jesus Christ), Pulchinello and San Gennaro on one arm, but Hamilton's attitude to the saint was laconic. Writing to his opposite number in Madrid, he assured him that nothing changed in Naples, neither court diversions nor popular distractions; 'St January performs his miracle in the usual manner, and at the usual Seasons.' Elsewhere he referred to the 'mixture of riot and bigotry' in Naples.

For Horace Walpole every political spasm in the city illustrated the symbiotic convulsions of nature and humanity. Echoing Voltaire, Walpole confessed to Hamilton that he regarded 'burning mountains as very petulant ovens, and a little destructive' and wishing 'you had not exchanged your taste in painting and antiquity for phenomena'. The convulsions of the earth were not for Walpole. He admitted to having no curiosity 'about the anatomy of Nature. I admire and revere, but am not more struck, probably less, with the dissection than with the superficies.' From the safe distance of Strawberry Hill, he hoped that lava might at least have aesthetic possibilities but Hamilton pointed out that once cooled it tended to crack, making it unsuitable for anything large scale. More to Walpole's taste was history painting, such as the work of a recent Neapolitan arrival, Angelica Kauffmann, who imagined the scene between Pliny the Younger as he steadfastly supported his mother at Misenum during the AD 79 eruption in which her brother had lost his life. However, the death of Catherine Hamilton in 1782 and the news, four years later, that William had become besotted with a voluptuous girl – Charles Greville's former mistress, who spent her time re-enacting poses borrowed from figures on Greek vases for the grandees of Naples – had opened a distance between Walpole and Hamilton.

Goethe, who was as gratifyingly seduced by Vesuvius as Walpole failed to be, spent some time in the company of Sir William and 'his Fair One', the increasingly notorious Emma, who was rarely drawn or painted (especially by George Romney) without the outline of the volcano behind her, even though she seems never to have made the effort to climb up to the crater. In the privacy of his journal, Goethe confessed that he found Emma 'frankly, a dull creature' with an unappealing voice and little to offer except her beautiful figure. In May 1787, returning after his excursion to Sicily, Goethe was ushered into the envoy's 'secret treasure vault' in the Palazzo Sessa. Wilhelm Tischbein, the German painter and art historian had described the aging Hamilton's 'lovely existence' in these rooms, where the subjects of the paintings ranged from boisterous children to Vesuvius in all its moods, an intensely personal choice, hung at random. Hamilton's collections told of his dedicated patronage of working artists, both Italian and foreign, his interest in minerals, lava and rocks, and particularly his eye for, and access to, the tide of newly excavated antiquities. Goethe judged Hamilton to collect everything from choice objects to chance bargains. In his view, his mistress reflected his tendency to buy anything lovely that fell his way, while Horace Walpole, referring to

Emma's repertoire of staged classical 'attitudes', said that he had merely married his gallery of statues.

✕

Hamilton's long tenure in Naples finally ended miserably and badly. Married to Emma since 1791, in late 1798 they and the royal family – together with their closest advisors such as Lord Acton, who had been a powerful figure at court since 1778 – were forced to flee from the French and their Neapolitan revolutionary allies, to the safety of Palermo during the short-lived Parthenopean Republic, during which Naples was held by Jacobin sympathisers. The king was soon reinstated and terrible reprisals taken. Hamilton's beloved Vesuvius was now a clichéd image of political revolution, both for and against. One 1793 cartoon shows the volcano titled the 'Torch of Truth' which spouts forth flames, annihilation, blasphemy, plunder, rape and murder.

In Sicily, Hamilton found himself once again within easy reach of Etna (compared to which he had considered Vesuvius to be a 'meer Mole hill' when he and Catherine toured Sicily in 1769) and in contact with passing connoisseurs such as the Earl of Elgin. His sojourn on the island was not a great hardship. While there he arranged for one of his favourite painters, Giovanni Battista Lusieri, to go to Constantinople and Athens with Elgin. But Sir William also had to find ways to settle his towering debts, founded upon the extravagances of three establishments but exacerbated by the British Government's habitual failure to remunerate its envoys. Back in London, by then an old, sick and derided cuckold (Emma and Nelson's daughter Horatia was born in 1801), he looked on at two major auctions of his collections, while another shipload was lost at sea. He died in spring 1803, leaving Vesuvius to the geologists, and Naples from 1805 to be ruled first by Napoleon's brother Joseph Bonaparte, and then by his brother-in-law Joachim Murat.

In his final report to the Royal Society on the 1794 eruption, Sir William Hamilton had quoted Seneca: 'we see what we are permitted to see, and reason as best we can.' The measured tone was a fair indicator of the man, for whom observation, not opinion, showed the way ahead.

4

....

ROMANTICS

If Vesuvius had previously served as a handy manual to the mysteries of the earth, now it offered a key to the complexities of the psyche. Vesuvius became a connecting thread running through European Romanticism. The suppressed violence of the volcano with its unpredictable (but increasingly regular) outbreaks, the source of which remained hidden and mysterious, could be seen as a metaphor for the conflicted soul as much as for revolution and radical political thought. Between each individual and the natural world now lay a vast range of possibilities, emotions to be triggered at will. The variety and complexity of responses Vesuvius evoked were, potentially, limitless. Even great literature was judged a distraction from pure experience. Mme de Staël counselled that visitors to Vesuvius avoid reading Virgil and Milton beforehand, for each response to the scene must be entirely personal.

The mood in the mid eighteenth century had tended towards 'a strong imagination' and it was the political theorist and philosopher Edmund Burke's ideas on the Sublime that drove the search to match the propositions in his text to reality. In his *Philosophical Enquiry into the Origin Of Our Ideas of the Sublime and Beautiful* (1757) Burke stated, 'Terror is in all cases … the ruling principle of the Sublime'. The aesthetic theory of the Sublime, calling up responses of awe shading through to absolute horror, was counterbalanced by the Beautiful, which called up peaceable emotions, engendered by the gentle pleasure of the scene or object at hand. The literary Romantics took something from both, while abhorring much about modern life, especially industrialisation.

The itinerary and terrain of the Grand Tour might have been designed to illustrate the Beautiful and the Sublime. The emerald meadows and blond stone chateaux along the Loire and, later, the wash of gilded evening light over the landscape of the Roman *campagna* epitomised the former. The travellers' Sublime experience began (or ended) with the Alps, Chamonix and its static river of ice – vertiginous bridges across rock gorges, torrents and waterfalls dashing from an immense height, and, for the fearless, a thrilling descent into damp caves crusted with supernatural-looking mineral growths. The southern counterpart to this, as fiery and unpredictable as the Swiss scene was frozen and awe-inspiring, was Vesuvius. In the summer of 1773 Voltaire – no Romantic, but father of the Enlightenment and arch-priest of the rational – wrote to thank Sir William Hamilton for a copy of his recently published volcanic observations but remained sceptical towards the superiority of the volcano, as an object in the landscape. He considered temperamental Vesuvius and Etna, so '*pleins de caprices*', to be like a pair of petulant little men whereas the Swiss mountains, as seen from his windows at Ferney in their immense scale and icy stillness, gave him a sense of '*un calme eternel*'. The eighty-year-old philosopher could not summon up any taste for unnecessary danger.

As if cued by Burke, little more than ten years after he had elaborated the notion of the Sublime, Vesuvius began a long period of regular activity. The scene was set. Almost a hundred years afterwards, the extremes of fire and ice, turbulence and stillness, were still inexorably linked – both metaphorically and figuratively, both in man and in landscape. The hero of the Reverend Thomas Gray's novel, *The Vestal*, the first of many melodramatic depictions of the final days of Pompeii, had a heart 'cold as Alpine snows [but] whose feelings were deep and intense as the secret fires of Vesuvius'. The prolific boys' adventure writer Captain Frederick Marryat, who climbed Vesuvius with his friend the scene painter and landscapist Clarkson Stanfield as it erupted on New Year's Day 1839, described himself dreaming at Niagara Falls: 'I wished myself a magician, that I might transport the falls to Italy, and pour their whole volume of waters into the crater of Mount Vesuvius; witness the terrible conflict between the contending elements, and create the largest steamboiler that ever entered into the imagination of man.' Yet despite his stock in trade of melodramatic subjects, Stansfield's sketches of Vesuvius in eruption have nothing superficially theatrical about them at all. He builds up huge, billowing clouds of grey vapour and ash, looking like grubby down-filled pillows, and then lets them smother the entire view, rather as if Marryat's dream had been his own.

⧗

Johann Wolfgang Goethe arrived in Naples in early 1787. His inclinations placed him neatly with one foot on the side of the Enlightenment, the other on that of Romanticism. Goethe was in search of the 'classic soil' but had also travelled to Italy to escape the constraints and difficulties of his life in Weimar. As a student in Freiburg he had attended lectures by the leading German figure in geology, A. G. Werner, who believed that the origins of the earth lay solely in the actions of the oceans. Goethe's preferred approach was a solid foundation of knowledge, and thus rationality, onto which he could graft his subjective response. His disappointment with the vaunted scenes of antiquity may have intensified his curiosity about, and susceptibility to, the phenomenon that was Vesuvius.

First he spent a day in the outskirts of Naples, marvelling at 'treacherous ground under a pure sky' in which 'barren and repulsive areas' gave way to natural luxuriance, even 'noble oaks' in an old crater. When he climbed a shrouded Vesuvius the following day, he discovered recent signs of activity, the latest lava flow just five days old. He was tempted to approach the crater but the dense fumes almost smothered him when the wind changed direction. His guide vanished smartly. In the letter in which he described this first ascent, he was quick to point out the scorch-mark on his paper. Evidence like the black trace of ash that fell onto the pages of Turner's sketchbook in 1819 (the year in which he finally arrived in Naples) was entirely visceral. To Goethe's intense excitement he realised that Vesuvius was coming to life. The intervals between the thunderous roars and the ejection of rocks, stones and dust, could be measured. Perhaps Sir William Hamilton had advised him how to read the mountain.

Soon Goethe decided to attempt Vesuvius again, despite the leaden negativity of his German *cicerone*, the painter Tischbein, who told him that he considered the volcano no more than 'a formidable, shapeless heap ... which again and again destroys itself and declares war on any sense of beauty.' Even Goethe's optimism was dented by the chaos of the traffic below and the dinginess of the 'realm of Pluto' (in this instance sited above ground) where, after a long period without rainfall, all the vegetation and every building surface were smothered in ash. Two guides hauled the men up to the plateau beneath the cone. Approaching the crater they were assaulted by a hail of different-sized stones from overhead as well as stumbling over those underfoot, and gloomy Tischbein 'saw that the monster, not content with being ugly, was now threatening to become dangerous as well.'

Goethe's mood improved dramatically as he sensed 'imminent danger ... [that] challenges Man's spirit of contradiction to defy it.' With the younger of their guides he decided to scramble up to the crater and dash back again in the interval calculated between the thunderous roar and the actual eruptive moment. As advised, he lined his hat with linen and silk handkerchiefs for protection and then, having grabbed the younger guide's belt for traction, found himself 'on the lip of the enormous mouth' though blinded by smoke and steam. They delayed, in the hope of seeing more, without counting the passing minutes. As they stood on a sharp edge of the crater, the entire mountain convulsed and 'a terrific charge flew past us'. They ducked and then ran, forgetting that now they had the advantage of another interval. Ash-covered on his return, Goethe was scolded by the heartily relieved Tischbein, who had waited under a rock. As innocuous distraction, they examined how the lava built up, layer upon layer, 'until finally the whole flow petrifies in jagged shapes' – like ice floes on a river.

In the following days Tischbein conscientiously showed Goethe around the galleries and took him to Pompeii, the 'mummified city' and then to Herculaneum and the Portici museum. Towards the end of the month Goethe heard that a new lava flow was in progress, this time to the north of the volcano, so out of sight from Naples. Taking the two guides he had engaged previously, Goethe (alone this time) walked as close as he could to the ten-feet-wide stream of lava, noting now how it cooled to either side and below, with the molten material in the centre deepening all the time. Scoria, the lightweight rock formed from recently solidified lava, rolled off the edges. He was even tempted to stand on the crust, 'twisted in coils like a soft mush', but immediately was pulled away from 'the hellish cauldron' and out of its suffocating fumes by the guide. Chastened, Goethe broke off specimens of lava, much as he had seen in the hands of the 'lava dealer', the condensed minerals showing clearly through the volcanic soot. The evening ended with a fine sunset. Goethe reeled in romantic, anaesthetised, confusion. 'The Terrible beside the Beautiful, the Beautiful beside the Terrible, cancel one another out and produce a feeling of indifference.'

His final evening in Naples before leaving for Sicily, Goethe dined at the royal palace in Portici. Someone flung open the shutters and there was Vesuvius, roaring away,

> and at each eruption the enormous pillar of smoke above it was rent asunder as if by lightning and in the glare, the separate clouds of vapour stood out in sculptured relief. From the summit to the sea ran a streak of molten lava and glowing vapour, but everywhere else sea, earth, rock and vegetation lay peaceful in the enchanting stillness of a fine evening, while the full moon rose from behind the mountain ridge.

His subjective response left his scientific interests lagging behind. No scene would ever provide him with such a combination of emotive thrill and spiritual calm, the emotions and the senses in complete equilibrium.

These extremes, the polarities between the forces of nature in the creation of the earth, reappear in Goethe's *Faust*, published between 1808 and 1829. Some choose to see the contemporary geological dispute at the heart of it, the devilish Mephistopheles representing the violence of the Plutonist theory in which a core of molten material was considered the formative element in the universe. 'Eternal fires were whirled in Earth's hot entrail'. The hero Faust, by this measure, stands for the calmer Neptunist view in which the earth has been created, rather more gently, by the action of the oceans. 'The hills with easy outlines downward moulded/ Till gently from their feet the vales unfolded!/ … Requiring no insane, convulsive changes'. The latter reading was where Goethe's sympathies would remain.

⧗

It was Milton's version of fiery demonic beings that came to the lively young traveller Katherine Wilmot's mind on her ascent of Vesuvius in 1802, as she took advantage of a lull in the long Napoleonic wars provided by the short-lived Peace of Amiens. Her outdoor adventure was an antidote to the claustrophobic hours that she had been obliged to spend at court. The Queen of the Two Sicilies, who had been Emma Hamilton's close friend, was 'a sturdy looking dame' but the King, with whom the wretched Sir William had spent such a large part of his life in Naples, no more than 'an overgrown ass' with an imbecilic expression. The heir to the throne danced 'like a cow cantering' and was so crude that 'vulgarity becomes genteel within his presence' as she put it, acerbically. Within a couple of years, they would all be dislodged (albeit temporarily), the throne taken by Napoleon's family instead.

It was a relief for Miss Wilmot to leave the depths of disintegrating Neapolitan high society and head upwards. Their climb, pulled up with handkerchiefs looped around straps and cords around the guides' waists, was tough going. Despite having specially strengthened boots, one of her feet was bleeding by the time they reached the top. At this point she and Mrs Derby, 'a beautiful little American', were expected to sit quietly, contemplating the stupendous view while the gentlemen pressed on and descended into the crater. But Martha Coffin Derby was among the earliest Grand Tourists of postrevolutionary America and neither she nor her doughty, witty Anglo-Irish companion Katherine were stereotypical sedentary ladies of rank.

Both women were recording their journey; Katherine noticed Martha 'writing a little letter to her friends in America, descriptive of the surrounding scenery.' In fact, Mrs Derby described them inching along a ridge 'with precipices on both sides where inevitable death must have been the consequence of falling'. They both shut their eyes against vertigo and Katherine's three guides 'hook'd my arm upon a pole, and dragg'd me along'. Then the intrepid pair descended into the hellish pit, going down through intense heat and smoke swirling from multiple fissures. At the very bottom the surface was 'dark green and yellowish' – from the sulphur, Katherine imagined – 'so broken into waves that it look'd like the sea in storm, suddenly congeal'd' and yet where the lava had already hardened into stone, green moss was nonchalantly flourishing.

It was there that Katherine Wilmot joked that they could have provided the ornamental element in Milton's *Pandemonium*, 'smoked as we were like demons, "prone on the ground extended long and large and floating many a rood"'. As she travelled through the sites of classical antiquity she grew angry that, in contrast to even her younger brother, her genteel education had entirely omitted to expose her to the great literature of the Greek and Roman periods. For Katherine Wilmot, with her confident enjoyment of seventeenth-century literature and acute eye, the climb down into the crater was a moment of pure freedom and a ringing riposte to those who had set handcuffs and blindfolds on her and her fellow women. Back at the resting place, the so-called Hermitage, they had difficulty convincing the 'good friar' that even the ladies in the party 'absolutely had been at the bottom of that terrible crater'.

Women, so circumscribed at home, gained a particular thrill from their journey into a landscape unlike any other in Europe. Georgiana, Duchess of Devonshire, banished to the Continent for her extra-marital affairs and illegitimate child (unlike her husband who remained unpunished for his) immersed herself in 'chemistry' and the study of minerals during her time in Naples, showed great flair for the subject and became a firm friend of Sir Charles Blagden, Secretary of the Royal Society. She returned home eager to continue her studies, bringing with her many samples of volcanic rocks and minerals, a fine collection which she continued to build, filling several display cabinets at Chatsworth, for study as much as aesthetic delight. The painter Samuel Palmer wrote from Italy that his new wife Anny was 'pleased beyond measure with anything volcanic and has the heart of a lion the moment she begins to slip about on cinders and ashes'.

When active, Vesuvius offered its devotees an array of testing physical and mental experiences, along with the visual stimuli and psychological promptings that the literary Romantics sought. The Gothic novelist, Ann Radcliffe, who witnessed the 1794 eruption, set her melodramatic novel *The Italian* on the Bay of Naples where the distant thundering and 'groans' of Vesuvius act as an all too literal counterpoint to the turbulent emotions of her hero, Monsieur Vivaldi. The Marquis de Sade who had climbed the volcano, with his valet (and confidant) Martin Quivo, in 1776, incorporated a wild and typically deviant scene on the crater's edge in his novel *Juliette*. Mme de Staël, through the mouthpiece of her fictional self, Corinne, the heroine of the eponymous novel, deduced from the erupting volcano the overwhelming power of nature, to penetrate the heart and to provoke true wonder. Man is not the source of the greatest mysteries, for 'an independent force threatens or protects him according to unfathomable laws', she wrote.

Mary Shelley began reading Mme de Staël's 1807 novel as she sat contentedly feeding her week-old child in their London lodgings in the spring of 1815. She continued to dip into it for several days, as she nursed the baby, but on 6 March she stopped for, as she wrote bleakly in her journal, she had found her baby dead. Amid her agonising, if telegraphic, record of grief, she was finding no relief in literature. The

Shelleys spent the summer of 1816 – the worst that Europeans remembered and caused by dust from the eruption of Mount Tambora in Indonesia the previous year – sitting gloomily on the shores of Lake Leman, Mary conjuring up the story and enduring image of Frankenstein.

It was not until mid December 1818, by now settled in Naples for the winter, that Mary Shelley took *Corinne* up again and quickly read it. Now perhaps she concurred with Lady Elizabeth Foster's view of the book:

> *I had become so attached to poor Corinne that I dared not read too quickly for fear of separating myself from her and for the fear I felt that you had destined her to be unhappy. How did you have the courage to make her suffer so? I liked Oswald but I cannot forgive him for Corinne's unhappiness. How well you have described the sorrows of the heart! You have painted them as faithfully as your eloquent descriptions of Italy.*

Mary's own situation was grim; her next child, one-year-old Clara had died in September, in Venice, and it seems highly likely that her step-sister Claire Clairmont was pregnant again, possibly even by her husband (Claire's first baby was Byron's). Percy Bysshe Shelley registered himself as father of a daughter in Naples on 27 December, the very day that Mary noted in her journal, 'Claire is not well'. Some argue that the child was their maid Elise's. The only certainty was that it could not be Mary's. Her creation was Frankenstein, a being born out of fire.

Corinne's emotions, mirrored by the extremes of beauty and misery of southern Italy where she was travelling with her English lover Oswald (who was soon to abandon her), were horribly apposite to the twenty-year-old Mary Shelley's situation. After she had finished reading *Corinne*, she handed it on to Percy. The following day, with Claire (who may have been within a few days of giving birth), they climbed Vesuvius. In a grim atmosphere of suspicion and unhappiness, the trio toiled up the mountain. The date was December 16th, the anniversary of the 1631 eruption, one of the three dates on which the phials of San Gennaro's blood were exposed each year, in the hope of liquefaction and the protection that the miracle would bring.

Shelley, like the eighteenth-century cartoonists, was fond of invoking volcanoes for their political message. In 1812, in his poem 'To the Republicans of North America', he pleaded with Cotopaxi, in Ecuador, to erupt with suitably revolutionary fervour. Five years later, in 'Marianne's Dream', he uses the convulsions of the earth, earthquakes, floods and volcanic eruptions, as metaphors for desperately needed political reform. Now, he was about to set foot on mainland Europe's only live volcano – a revolutionary Romantic experiencing the ultimate in Sublime aesthetics. He was alert to Vesuvius as a prompt to his own psyche but his companions' responses seemed a matter of indifference to him.

Afterwards, Shelley wrote a long letter to Thomas Love Peacock. He began by comparing the 'deformation and degradation of humanity' of Neapolitan life with the beauty of the landscape and the gentleness of the climate, and described their recent journeys in the footsteps of the ancients. They had set out – no doubt quite gently if Claire was in the advanced stages of pregnancy – with a trip to the Elysian Fields, which had been a disappointment. Close reading of Virgil had unduly raised their expectations. Then they took a boat along the Bay of Pozzuoli and up to Baiae, landing to visit Lake Avernus. Nearby Monte Nuovo offered a vivid reminder of how volcanic action could transform the landscape virtually overnight but, surprisingly enough, Shelley thought that the Solfatara was 'not a very curious place' despite Petronius' account which should have prepared him for a chaotic, barren place 'watered by the streams of hell', from which 'grim Pluto from the ground reared his dire form, while played around his head, with smouldering ashes strewed, sepulchral fires.' Perhaps they should have heeded Mme de Staël's caution against too much literary preparation.

Then came the day allocated for their ascent of Vesuvius. At Resina, Mary and Shelley took mules but Claire continued the journey by chair, mounted upon the shoulders of four men 'much like a member of Parliament after he has gained his election, and looking, with less reason, quite as frightened.' At the Hermitage of San Salvador they ate, courtesy of the usual worldly recluse whose robe was tied with rope, monk-fashion.

Apart from the glaciers they had seen earlier, Vesuvius offered 'the most impressive exhibition of the energies of nature I ever saw', Shelley told Peacock. Lacking their immensity, magnificence or even radiant beauty, it was Vesuvius' 'character of tremendous and irresistible strength' that proved so potent. Unsurprisingly enough, Shelley's reaction was symmetrically opposite to Voltaire's.

As they climbed the mountainside, the lava, bumpy under the feet of their mules, seemed like 'the waves of the sea, changed into hard block by enchantment'. It was, he added, 'difficult to believe that the billows … are not actually in motion'. And yet this strange static flood had quite recently been a 'sea of liquid fire'. Beyond the Hermitage, they continued on foot. Shelley thought the difficulties had been greatly exaggerated; by negotiating the rocks (lava) and ash-filled hollows with care, it was not particularly tiring. At the summit, came a plateau of 'the most horrible chaos that can be imagined; riven into ghastly chasms, and heaped up with tumuli of great stones and cinders, and enormous rocks blackened and calcined', all expelled violently from the volcano and piled everywhere. In the centre of this hellish landscape was the cone, currently in a 'slight state of eruption' with billowing white smoke, columns of dense black vapour out of which fell hot embers and black ash, against a soundtrack of deep, subterranean rumbles. It was a fittingly cathartic auditorium in which the author of *Prometheus Unbound* could watch the theatre of Nature.

They were sitting close enough to be showered with ash; the activity within the volcano was giving their expedition a satisfactory dash of danger. On every side, rivers of lava rolled out, crackling as they crept along, except for one which fell over a precipice, 'a cataract of quivering fire'. Gingerly they approached a lava stream; Shelley judged it to be about twenty feet wide and about ten deep, the intense heat betrayed by a haze on the air and the rolling sulphurous smoke. They stayed, as every visitor was bound to do, to witness the sunset before the night revealed 'streams and cataracts of the red and radiant fire', and overhead the white heat of ejected rocks etched vapour trails onto dense black cloud.

The intensity of the experience had made Shelley entirely self-absorbed, insensitive to the feelings of others. Mary Shelley's journal entry tells its own story. She telescopes the day brutally: 'Wednesday, Dec. 16. – Go up Vesuvius, and see the rivers of lava gush from its sides. We are very much fatigued, and Shelley is very ill. Return at 10 o'clock.' By the time they were back at the Hermitage, guided down by men with flaring torches, Shelley was in acute pain (although he did not mention it to Peacock in his own account). The women became desperately worried. There was no reassurance at hand; the vile behaviour of the guides ('native pirates' as Mark Twain called them) lived up to Shelley's worst predictions, in particular the men who carried Claire's palanquin who had threatened to leave her in the middle of the road at night. But Shelley still admired their startling physical beauty and was enchanted when at nightfall, 'they unexpectedly begin to sing in chorus some fragments of their wild but sweet national music'.

The ignoble savages, showing occasional signs of native nobility, were by now among the established romantic clichés of Vesuvius and had become a kind of *commedia del'arte*, as much a semi-fictional link between past and present as the liquefaction and miraculous powers of San Gennaro. Everything helped to build up an image of exotic culture, colourful and inherently primitive – visitors to southern Italy were transported back to the realms of, respectively, folklore and idolatry, an uncomfortable paradox to those who put their faith in classical literature. But at least in Pompeii the population could no longer trouble the visitor. Even Mary was happier now; 'Go to Pompeii. We are delighted with this ancient city.'

For Shelley, composing *Prometheus Unbound* in Rome the following spring, in the delirious aftermath of his near-breakdown in Naples, he wove the imagery of the remembered volcano into that of the Baths of Caracalla, so readily to hand and of which he wrote, 'Never was any desolation more sublime and lovely'. In the poem, the polarities of the Romantic landscape, fire and ice, gave way to Demogorgon's underground lair – in other words, Vesuvius – from which would come with eruptive force, new energies to deal with tyranny, even including electricity, newest of all fires. Having completed the poem, he went to see the spectacular Easter girandoles at St Peter's with Mary and Claire.

⌛

Shelley's interests may not have included the science of vulcanology, but the young resident Professor of Chemistry at the new Royal Institution (founded in 1799), Humphry Davy, combined his science with a dash of Romantic poetry – his own. Currently he was extracting entertainment from the abstruse topic of volcanic eruptions, having choreographed a small one to take place on the table between himself and the packed tiers of fashionable men and women in the lecture theatre on Albemarle Street. The meticulous Michael Faraday took notes. Davy told them that he believed the Earth's interior to consist of earth metals and alkali. If concentrated and combined with oxygen they were 'a very complete and a very probable solution to the Phenomenon of Earthquakes and Volcanoes and perhaps considered thus they may lay the foundations of a new and perfect System of Geology'. The chemical explanation to which he subscribed had been current for over a century but Davy was stronger on analysis than his forebears. He presented 'a little arrangement that will help to illustrate what I have said … a very feeble exhibition but it will convey an idea of my meaning. It is a small volcano formed of clay &c. in the shape of a mountain and having two or three pieces of the Alkaline metals introduced here and there.' By adding water to his volcano, 'I shall be able to inflame it and cause it to burn briskly.'

Davy reassured his rapt audience that the fire would not last long but it burns 'somewhat in the manner that I suppose Volcanoes do'. Faraday drew a sketch of the tiny volcano, a pile of earth and stones in mountain shape, erected on a square board. Meticulously he described the performance: 'Certain openings and fissures were left in the top and sides of this mountain in which was put pieces of the metal Potassium. Sir H Davy took a bottle of water and poured some of its contents into the fissures. Inflammation immediately took place, light, heat and smoke was emitted and the little volcano vomited forth flames.' It was complete in all but 'the production of ashes and lava', obviously not achievable on such a small scale. In fact, long before that, the eminent seventeenth-century French chemist Nicholas Lémery, a believer in demonstration over theory, carried out his own volcanic experiment. Erring on the side of caution, the so-called *volcan de Lémery* was buried. It consisted of a paste of iron filings and sulphur to which water was added to set off the explosion.

Davy's hyperbolic biographer, John Paris, also reported the experiment at the Royal Institution. As he put it, water activated the metals and all was 'thrown into violent action – successive explosions followed – red hot lava was seen flowing down its sides, from a crater in miniature – mimic lightnings played around, and in the instant of dramatic illusion, the tumultuous applause and continued cheering of the audience might almost have been regarded as the shouts of alarmed fugitives of Herculaneum and Pompeii.' Paris could

14. Sir Humphry Davy portrayed by Thomas Phillips in 1821, in the pose of Regency poet rather than man of science.

turn anything into a melodrama but Davy's emphasis was primarily on science, even if he was in the literary circle of Robert Southey and Samuel Taylor Coleridge. The latter had gained firsthand volcanic experience well before his friend. Coleridge climbed Etna twice while in Sicily in 1804, and was haunted by the pitchy depths of the crater long after. The following year he was in Naples and visited Vesuvius, which he had evoked, in his poem 'The Nose' of 1789; like Pliny, 'I perish in the blaze while I the blaze admire.'

Still in London, Davy was tussling with the origins and characteristics of volcanic matter from the perspective of a chemist, using electrolysis to isolate the various metals. In 1807 he became a founding member of the Geological Society. Davy had a foot to either side of a growing divide. Southey pleaded with his chemist friend not to lose 'the habit of seeing all things with a poet's eye', while Coleridge, for whom Davy secured a series of lectures on literature at the Royal Institution in 1808, promised Davy that once he had sorted out his own affairs he would 'attack chemistry like a shark'. In a poem at this time Davy made the point that poetry and science were joining forces to probe the universe, as a result of which 'the renovated forms/ Of long-forgotten things rise again.' Yet in his notes for the lecture on volcanoes he accepted that it was impossible to do full justice to a 'combination of circumstances in which feeling & hearing & sight are almost equally concerned.' Vesuvius served the purposes of a wide, and widening, church in which no two people saw it in the same light or through the same lens.

Accompanied by Faraday, Davy's first visit to Vesuvius itself was in May 1814; they walked up from Resina, enjoying and noting the richness of the soil and vegetation, and arrived at the Hermitage where the black-gowned recluse 'proved himself not all deficient in the art of an innkeeper'. The usual *opera buffa* between bemused visitors and rascally guides then took place. But Davy and Faraday were not to be distracted. The edge of the crater gave them a clear view into 'an enormous funnel', encrusted with a yellow substance that Davy identified as iron chloride. As so often, the wind direction suddenly changed and they retreated at speed, pursued by a cloud of 'sulphurous acid gas'; Faraday had lost time collecting mineral samples and only just escaped over burning lava 'to the great danger of my legs'. They admired the polychrome deposits and watched their eggs being boiled on the lava before wolfing them down with bread and wine. After that, they slithered down through the ash, a rapid but unpleasant descent, only to repeat the entire excursion the following day – this time to see it by night.

That evening was the 14th May. Even by the popular standards of the time, an abnormally large crowd climbed the mountain as darkness fell and 'the flames appeared more and more awful – at one time enclosed in the smoke, and everything hid from our eyes; and then the flames flashing upwards and lighting through the cloud' before the wind blew and the sky cleared, revealing the flames leaping high out of the mouth of the crater. To this undeniably Sublime scene was added a different twist – for that one night. The guides laid cloths down on the lava, and under flaring torches set out a feast of chicken, turkey, cheese and wine as well as the inevitable boiled eggs for a bevy of different nationalities. After the meal was over, a toast was offered to 'Old England' and both 'God save the King' and 'Rule, Britannia' were sung, followed by two 'very entertaining Russian songs' – strange and touching music, Faraday thought. The lavish meal and the celebrations marked Napoleon's defeat and quite possibly the end of Joachim Murat's rule over Naples. The euphoric picnic on the fiery flanks of the volcano marked a moment that even Shelley might have appreciated – since he considered Napoleon a mean-spirited *'unambitious* tyrant'. But for once, Vesuvius rose above metaphor.

By next March, when Davy and Faraday were back in Naples, Bonaparte had escaped from Elba. On this occasion Davy planned to examine Monte Somma. The men did not linger for hospitality at the Hermitage but carried on, while their guide continued his tired recitation of great eruptions of the previous century and the miraculous powers of San Gennaro. Faraday noted, drily, that the image or saint sometimes fell short of expectations and that 'even the Virgin Mary has been so much abused as to have phlegm thrown in her face.' Faraday then continued towards Vesuvius, considerably altered since their previous visit. At the crater's lip, his guide urged him even closer, the two leaping from rock to rock. At one point Faraday slipped and looking up saw a hail of red embers coming down and felt the ground heaving below. He regained his footing just in time and they gazed into the apparently bottomless volcano. 'Here the scene surpassed everything.' Great muddy 'splashes' of lava were expelled, often splitting in the air and on landing, remaining red hot for up to fifteen minutes. After a particularly menacing explosion the guide cautioned him to move away, lest the crater edge break off, and so he 'retreated a little and then sat down listened and looked'.

Back in Naples, Faraday was calmly satisfied after his alarming experience. 'I have gained a much clearer idea of a volcano than I before possessed.' Meanwhile, in May 1815 Murat was deposed in favour of the disappointing monarch Hamilton had known so well. Joseph Franque painted the restored Ferdinand IV standing awkwardly on the slopes of Vesuvius, surrounded by his military entourage in resplendent uniforms and insignia, while the queen sits uncomfortably in her conveyance. Unusually, the volcano has become a symbol of stability and continuity.

Davy, whose observations on Monte Somma were regularly reported to the Royal Society, was consciously travelling in the footsteps of Sir William Hamilton, to whom he was indebted, he said, for bringing accuracy and his 'elegant and acute mind' to the study of the mystifying phenomenon of volcanic behaviour. In that spirit, on the basis of his observations on the spot, Davy entirely revised his ideas. He paid further visits to Vesuvius in 1819–20 and was fondly remembered by the chief guide, Salvatore, as the writer Anna Jameson was told. Ever a chemist, Davy also attempted to provide an explanation for the liquefaction of the blood of San Gennaro but eventually concurred that it was miraculous. He also attempted to find a means to unroll the scrolls of Herculaneum – those papyri that had defeated poor old Padre Piaggio.

In 1828, the year before his death, Davy finally read his *On the Phenomena of Volcanoes* to the Royal Society. He had changed his mind gracefully, no longer believing that volcanic activity was caused by chemical reaction. He accepted that the evidence of intense heat within the earth offered 'a still more simple solution of the phenomena of volcanic fires than that which has just been developed.' His (poetic) *Consolations of Travel* allowed for his confession, though in the words of one of three philosophizing protagonists: 'For a long while I thought volcanic eruptions were owing to chemical agencies of the newly discovered metals of the earths and alkalis, and I made many, and some dangerous, experiments in the hope of confirming this notion, but in vain.' Davy was also troubled by the conflict between the Christian view and the conflicting evidence in the natural world. The three philosophers, one himself, the other pair an enlightened Roman Catholic and a sceptic, discuss their difficulties. At one point they argue on the lip of Vesuvius, the contrast between the desolation after a recent eruption and the glories of the landscape below beautifully evoked. Yet for all Davy's strenuous attempts to align the three diverging strands – Romanticism, science and theology – they were already parting company.

The Romantic emphasis was on mood rather than spectacle, on individual response rather than pleasing the crowds. When the Norwegian Johan Christian Dahl painted Vesuvius from Quisisana, he showed the volcano smoking peaceably in the middle distance in the late evening light. He chose to frame it through the window, somehow guaranteeing the veracity of this unorthodox, asymmetrical view. Yet Dahl could also turn his hand to old-fashioned, somewhat clichéd views of Vesuvius erupting, set against the moonlight pouring over the water like an oil spill (the 'frozen music' of northern Romantics). Travellers' portraits too had become contemplative rather than bombastic, the subject seen in a quiet interior or a peaceful glade, holding no more than a book, rather than the theatrical repertoire of classical loggias and libraries, with antique busts and blue-blooded dogs.

Franz Ludwig Catel accompanied the neoclassical architect Karl Friedrich Schinkel to southern Italy in 1824 and captured his friend sitting by another open window, on another golden evening, beyond which distant Vesuvius emits no more than a wisp of steam. Catel carefully portrays Schinkel in a setting well-suited to the eminent architect's restrained style, Romantic Classicism, and temperamentally in tune with his northern sensibility. In fact, Schinkel had seen Vesuvius long before, in 1804, and came home to paint 'Vesuvius in its Infernal Fire-Glow', one of a series of 'perspective optical' landscapes (with music) for a public panorama in Berlin in 1808. He had even painted a stage set focused on Vesuvius. By the 1820s his own maturity, and the changed times, brought a very different perspective on southern Italy. Similarly, in 1819 Turner's first view of Vesuvius was a transforming experience, set against his previous set pieces of erupting volcanoes, Mount Vesuvius and Mount Soufrière (to which he added a cod-romantic poem). Turner's ethereally beautiful watercolours of Vesuvius, calm across the Bay of Naples, were in the greatest possible contrast to his earlier efforts. Where possible, Samuel Palmer wrote, every landscape painter should visit Italy for 'it enlarges his IDEA of creation'.

Opera fed off the turmoil and the destruction of volcanic eruption, both for scenario and sets. In 1815, Schinkel was the designer for *The Magic Flute* and painted Vesuvius dominating the backdrop for Act 1, which is set in the Egyptian Temple of Isis at Pompeii, newly discovered when Mozart and his father had visited the site. More spectacular sets still graced the first staging of *L'Ultimo Giorno di Pompeii* by the Sicilian composer Giovanni Pacini, performed at the Teatro San Carlo in Naples in 1825. Daniel Auber's *La Muette de*

15. An elaborate reenactment of *The Last Days of Pompeii*, an extravaganza in fireworks by the Pain brothers of London, was a popular late nineteenth-century attraction at Manhattan Beach, Coney Island, described as a 'marvellously realistic reproduction of an historic catastrophe' and a 'colossal spectacle'. Decades after the publication of Edward Bulwer Lytton's novel the choice of subject matter for these pyrotechnics pulled in vast crowds throughout the summer season.

Portici (also known as *Masaniello*) was premiered in 1828 in Paris, on a stage dominated by the moods of Vesuvius. Painters of theatrical tendency such as Francis Danby, Philip de Loutherbourg and John Martin extracted the maximum melodrama from the volcano. But few lavish opera productions nor scene-stealing paintings could compete with Edward Bulwer Lytton's novel *The Last Days of Pompeii*, a histrionic book that helped to launch an entire genre. Academic painters threw themselves at the subject, inspired. Every convention from the life class appeared in exaggerated form on canvas; anatomical complexity and frenetic incident combined to dizzying effect. In cities all over Europe crowds stood mesmerised before hyperrealist compositions based on the destruction of Pompeii, such as that by the Russian Karl Briullov, which predates Bulwer Lytton's novel and may have been his own inspiration, or the Frenchman, Henri-Frédéric Schopin. Doomed glamour, as Jonathan Keates neatly terms it, became the mood of the moment.

Bulwer Lytton wrote *The Last Days* while he was in Naples over the winter of 1833–34, where he and his wife had gone to repair his health (successfully) and their marriage (unsuccessfully). As so often, Vesuvius was the backdrop to personal catastrophe. (In Victorian 'sensation' novels of the 1860s, incendiary family relationships were often described as volcanic.) Bulwer Lytton depended upon Pliny the Younger's account for the physical details of the eruption itself, even though he varnished and tinted the picture with splashes of dazzling colour. After introducing the calamitous spectre of Pliny's pine-shaped cloud – 'the trunk, blackness, the branches, fire!' – he described, with mounting hysteria, 'that awful shower! … a sudden and more ghastly Night rushing upon the realm of Noon', only broken by lightning flashes with 'their horrible beauty' bringing 'their varying and prodigal dyes' to the scene. The catastrophe in his fictional foreground had a suitably figurative background. In August 1834, as the book was published, Vesuvius erupted once again. Immediately opportunistically adapted for the stage in New York (by Louisa Medina, an Englishwoman) *The Last Days of Pompeii* was packing audiences into the Bowery Theatre six months later and broke all the records for the length of its run. In Bulwer Lytton's opulent new house in London's Mayfair, he proudly presided over a splendid Pompeian chamber. At its heart was a perfumed pastille burner in the shape of Vesuvius which burned merrily away to beguile his guests.

Yet the persistence of what Nelson Moe calls 'imaginative geography' led to the determination by a new school of Italian realists, writers and artists, to erase the picturesque in favour of the actual. As early as 1834, Gaetano Forte painted Vesuvius as the distant, passive background to a working, semi-industrial set of buildings in the foreground, the outskirts of a modern city. The scene could have been anywhere in the north, Milan or Bologna, and the volcano merely serves to determine that the city in question is Naples.

The Italian Romantic poet and thinker, Giacomo Leopardi, grappled with the disparities and the misconceptions about the extremes of his own country too, between the north where he was born, and the south where he died, tragically young. The distorted hyperbolic picture of the latter, as held by northern Europeans, enraged and infuriated him. Rather than extolling the 'southern difference', the romantic and folkloric view that gained popularity later in the nineteenth century, Leopardi wrote that viewed from here, a villa at the foot of Vesuvius, the human race evidently had no more than a fragile hold on the earth, like the ant (Voltaire's comparison, too). Written in 1836, Giacomo Leopardi's *La Ginestra* proved to be his lengthy epitaph (he died in his late thirties the following year) as well as a stirring political and personal manifesto.

As Moe argues, Vesuvius stands for Leopardi's own harshly revised attitudes, turned against the comforting certainties and homilies associated with the panacea of progress. 'Sterminator Vesevo' (Vesuvius the exterminator) is simply the force of Nature. No metaphor could be more apt or resonant than mercurial, unpredictable Vesuvius with its subtext of death and burial.

Subverting a picturesque vision by the rawness of the terrain, Leopardi envisaged the volcano as a bleak threat, a natural destructive force primed to endlessly repeat its cycle. That image gave the lie to a progressive, modernising view of the future and dealt a blow to any incipient triumphalism stirring in the north. The evident capacity of natural force to eradicate everything in its path was sobering. If society in this 'vain and fatuous century' believed that it was seeing the advantages of progress, then it was time to pause and consider. Vesuvius was authentic, unarguable, however uncomfortable and deadly it might be.

Previously, passing literary and artist Romantics had requisitioned Vesuvius as the palimpsest of their various struggles, but Leopardi made it a warning symbol, having vested it (in the most general sense) with profound political and moral doubts about his country and his epoch. Soon after, the lengthy, messy struggles known as the Risorgimento (the Resurgence) would succeed in binding the disparate elements of the Italian peninsula into a single, if awkward, political and administrative entity. The Kingdom of the Two Sicilies was gone by 1860.

16. Portable and inexpensive, early souvenirs of Vesuvius often played upon the notion of heat. Ornamental fans, pastille burners and snuff-boxes were all decorated with images of Vesuvius, usually giving off the obligatory puff of smoke, as here, suitable keep-sakes for the less affluent or already overloaded Grand Tourist.

5

···

MAKING VESUVIUS

Such was the fever for excitement in the landscape, it was bound to lead to extremes. The Georgian architect and landscape designer Sir William Chambers argued convincingly for the introduction of Sublime effects, what he knew to call 'scenes of terror' after Edmund Burke's prescription, drawing upon his own youthful visit to China. He claimed that the Chinese concealed, high on mountain peaks, 'founderies, lime-kilns, and glass-works' in order that flame and thick smoke might issue forth and 'give to these mountains the appearances of volcanoes'. This was landscape gardening *con brio*.

His German friend and follower, Friedrich Wilhelm von Erdmannsdorff, took his admiration even further than probably even Chambers had intended. For many years he acted as the right hand to Leopold III, Friedrich Franz of Anhalt-Dessau, known as Prince Franz, helping him to achieve an extraordinary fusion and range of architecture and landscape design at his Wörlitz estate, the idyllic setting for the Prince's wider Enlightenment scheme of religious tolerance, as well as political, educational and agricultural reform. After a visit to England in the early 1760s, the pair turned south to undertake a more conventional Grand Tour and, after a period in Rome under Winckelmann's guidance, they visited William

Hamilton and climbed Vesuvius in 1766. Von Erdmannsdorff was back in Naples in 1790 while the prince's estranged wife Luise made her own Grand Tour in the late 1780s.

The Schloss Wörlitz was to be Prince Franz's principal seat, built in the 1770s and the first neoclassical country house in Germany. Throughout the park, intersected by a complex system of lakes, canals and watercourses, largely laid out as a naturalistic *englische garten,* were scattered dozens of structures, bridges, eye catchers and even a garden temple doubling as the town synagogue. There was a Chinese garden, complete with a pagoda, very much after Chambers' taste and model, in the grounds of the Oranienbaum, an earlier country house nearer Dessau. At Wörlitz, the Prince rebuilt the Protestant church in the town in a light Gothic style, echoing the style of his private refuge which was closely modelled on Strawberry Hill, the idiosyncratic creation of Hamilton's cousin Horace Walpole.

Unlike the rest of the buildings at Wörlitz, nothing remotely like the German Vesuvius existed in any of the parks and great landscape gardens that the Prince and his architect had visited and admired so greatly during their visit to England. The displaced volcano sat upon its own island, mirrored in a body of water configured to be the Bay of Naples. Built between 1788 and 1794, the Vesuvius at Wörlitz was considerably larger and more convincingly three-dimensional than any of the model volcanoes that were in vogue as public entertainment in European capitals. Then, as now, the volcano proudly, if oddly, sat looking at its reflection in the miniature lake, sitting like an incongruous ant-hill in the pancake-flat landscape of the region. It was deficient in just one particular – it never took its audience by surprise.

17. The lake at Wörlitz reflects the artificial volcano on the 'Stein' as it erupts, thrillingly, in 1795. Visitors view it from a 'gondola'. The basalt columns can be seen to the left, on the water's edge and tufa-effect rockwork adds to the feeling of being in the actual foothills of a volcanic mountain. Only the Villa Hamilton cannot be seen, being to the rear of the island.

The rugged tufa pile, roughly mountain-shaped, surmounts a labyrinth of grottoes and themed rooms (including one for night and one for day) and, when fired up from a central cell, becomes a pyrotechnic cocktail. Among the ingredients of the Sublime experience as defined by Burke were the shock of suddenness, dramatic sound – 'excessive loudness alone is sufficient to overpower the soul' – and the contrast to be found between brilliant light which 'as it overpowers the senses is a very great idea'; and utter darkness 'more productive of sublime ideas than light.' A volcanic eruption, even when manufactured to order, had every ingredient that the philosopher ordained. It left none of the senses untouched and was an exquisite illustration of Burke's central argument that 'opposite extremes operate equally in favour of the sublime, which in all things abhors mediocrity.'

At Wörlitz, eruptions were announced by a menacing rumble emanating from deep inside the miniature cone, accompanied by the putrid but suggestive smell of sulphur. Water was pumped up to the summit where it was discharged to cascade down over a hidden red lamp. Seen from a distance it looked convincingly like a moving tongue of red-hot lava.

Below the volcano sat the so-called Villa Hamilton, closely modelled on the envoy's Casino on the beach at Posillipo, the decorative theme of its three little rooms a celebration of the Bay of Naples ancient and modern. On a lava-topped table inside lay a copy of the *Campi Phlegraei* – a gift from Franz's estranged wife. Near by were some basalt columns, leaning at an angle, as at the Giant's Causeway, his nod to the great geological controversy of the moment. The entire elaborate conceit was the Prince and his architect's homage to the British envoy, personally known to both himself and von Erdmannsdorff, as well as to their compatriots such as Winckelmann and Goethe. The latter was a regular visitor to the Prince's rustic empire of enlightenment, considering it 'a fairy tale' and yet his novel *Elective Affinities* sets his wealthy, bored characters strenuously, and pointlessly, replanting a landscaped park, remodelling the paths and buildings, as if to distract themselves from their torn loyalties and to stave off impending emotional disaster.

Just before the completion of the Stein, the artificial island crowned by an imitation Vesuvius, Princess Luise had returned from southern Italy where her guide was the landscape painter Philipp Hackert (Goethe's drawing master). As a wronged wife of firm views, she refused the opportunity to meet the new Lady Hamilton but happily spent time with Sir William. Everything suggests that she was, at the very least, the joint instigator of the Vesuvius project at Wörlitz. Pompeian and Vesuvian motifs appear in several rooms of the Schloss but von Erdmannsdorff introduced a still more comprehensive ornamental scheme in Luise's own delightful villa, the Luisium, including four views of Vesuvius in eruption (scaled-up copies of Pietro Fabris' plates) at frieze level in the Pompeian Cabinet, a charming little room overlooking the park, also planted in the English style. In the Gothic House – a love-nest for himself and the gardener's daughter (another Luise, though many years his wife's junior) – is a special indoor seat from which to view the volcano. Sir William Hamilton had appropriated Vesuvius through knowledge, but Prince Franz appropriated it physically.

When the Scottish gardener, John Claudius Loudon, described all this in his *Encyclopaedia of Gardening* (1822), he managed to make it sound like Vesuvius itself. 'When the volcano is working, all kinds of inflammable materials are burned; when an immense smoke issues from the numerous apertures, and covers the top of the mountain with heavy black clouds. At the same time millions of sparks, rising from the gulf, form columns of fire, and streams of melted lava appear to flow down the sides of the mountain.'

No one else went so far as to build themselves a *faux* volcano but the imagery was potent. Tufa and pumice stone were perfect lightweight materials for garden structures and eighteenth-century landscape gardens were liberally peppered with grottoes, fountains and rockwork. Later on, came a fashion for quarry gardens, like the nineteenth-century public park, the Buttes de Chaumont in Paris, but nobody built a volcano again. Flaming urns evoked everlasting life on funerary monuments, but the flames were stonily unchanging.

⌛

The playful setting of these landscapes fed an appetite for festivities; elaborate firework displays over water became *de rigeur* for the wealthy, the night skies over their lakes and great gardens painted with choreographed light effects and explosive sound. For those without the space (or money) to build follies, hold fabulous parties or travel abroad for months at a time, theatre played its part in bringing the experience of far places and magic atmospheres close to home. By the late eighteenth century, by which time pleasurable excursions to continental Europe were on hold due to the Napoleonic Wars, a range of hair-raising effects such as sheer precipices and dank caves, wild weather and spectacular light effects, together with famous buildings and dramatic historic or biblical episodes, provided a heady tipple of blended catastrophe and shock for play-goers. Representations of fire were always firm favourites. There could be no more fearsome scene than a volcano in eruption and the ultimate example was Vesuvius, obligingly by then intermittently, but fairly regularly, active. Burke's esoteric aesthetic definition had entered common usage.

When he arrived in London in 1769, the Dublin-born scene painter Robert Carver lost no time in producing two contrasting scenes of Mount Vesuvius in eruption for the Theatre Royal Drury Lane: one seen distantly over the Bay of Naples with the waters offering a pleasing reflection (and double the effect) and the other, in fearsome detail, illustrating the lava flow, 'like a river of liquid fire, falling into a cascade from a rock'. The mention of 'New scenes' on a theatre poster guaranteed a full house. Philip de Loutherbourg, who came to London in 1771, via Strasbourg and Paris, also worked for David Garrick. His repertoire, shared between his stage sets for Drury Lane and his immense canvases destined for the walls of the Royal Academy, included thrilling avalanches and horrid sea battles, flaming iron foundries seen by night and menacing lamp-lit prison cellars. He was expert in the use of backlit, strongly-coloured transparencies and masterly at dramatic effects in paint. Shadow and silhouette played their part. With his little Eidophusikon (a word combining the Greek for nature and image), a kind of chamber theatre based upon illusion on a small scale, de Loutherbourg took the theme further still. No longer was there any need to seek plays whose mood might suit the scenery, for the scenery itself had become the point. Seats were always full to watch the erupting volcano at his establishment in Lisle Street, off London's Leicester Square.

The Marylebone Gardens had long been holding elaborate firework displays and for the king's birthday in 1772 they commissioned the former pyrotechnician from Versailles to design an entertainment, part theatre, part scenic display, called 'The Forge of Vulcan' with additional cast members including the Cyclops, Venus and Cupid. The dénouement was a fine volcanic eruption. This was faithfully copied at Ranelagh Gardens in Chelsea, with music by Haydn, Handel and others. Although the volcano in question was Etna, the promoters were becoming practised at the special effects required, starting with thick smoke, flaming craters and streams of lava.

It was the emphasis on visceral thrill that prompted public volcanic exhibitions in their various forms and on wildly different scales. They ranged from miniature cork models (including one so realistic that it set the establishment alight with 'volcanic' embers) to full-scale three-dimensional stage sets. Hamilton's invention was a pioneering moving image (surely Garrick had shown his to de Loutherbourg?) and now a procession of increasingly complex devices appeared before the public, mostly different versions of peepshows and magic lanterns. One depended on rotating a pair of glass slides in opposite directions with a handle, which gave alternating or dissolving effects in what is called a 'rackwork' mechanism. The first image might show a blackened crater, the next a burning vortex of fire and flame.

Benjamin Silliman (a professor of geology and mineralogy from Yale) witnessed a volcanic entertainment in London in the early 1800s. He was taken behind a curtain into a darkened room at Dubourg's central London showroom to witness a very credible eruption involving a small cork mountain, complete with live glowing embers, impressive lava effects, a repertoire of sound effects and – another advance on Hamilton's device – even the convincing smell of sulphur 'and such other effluvia as volcanoes usually emit'. Soon even the Royal Institution followed suit when Humphry Davy conjured up his table-top volcano.

In Britain, volcanoes were erupting nightly in cities from Glasgow and Manchester to Dublin and London – in special halls, opera houses, theatres and pleasure gardens, taking their rumblings, fiery effusions and sulphurous smells into some unlikely places. The dizzying excitement that affluent travellers might have experienced on Vesuvius or Etna became everyman's popular currency and these entertainments spread round the world, even as far afield as Australia.

John Burford's *Ruins of the City of Pompeii* was a painted panorama displayed in 1824 within his custom-built hall in central London, in which ticket holders enjoyed the scenery wrapping around them, providing a 360-degree view. A key to the features on view was provided. Soon Burford sold it to Frederick Catherwood who had a similar rotunda in New York. However, the public quickly tired of static exhibits and promoters introduced rolling displays (with the advantage that the spectacle could be transported to smaller venues and ones far afield) or dioramas with an ever more sophisticated range of supplementary effects. London's Regent's Park Diorama opened in 1823 just a year after Daguerre had invented it in Paris. The full panoply of the latest stage mechanics – lighting, sound effects, projected colour transparencies and more – was in play. Visitors could watch separate episodes from revolving platforms. Panoramas could be combined with 'pictorial' or operatic music to even greater effect; in 1844 the entrepreneur and musician Louis-Antoine Jullien put on his 'descriptive fantasia' at Drury Lane called 'The Destruction of Pompeii'. By 1845 even the citizens of Mexico City could share a *View of Naples with an Eruption of Mount Vesuvius* in Daguerre's Gran Diorama.

Against all this, outdoor re-enactments of Vesuvius were becoming formulaic. What had once been a spectacular night out was now an over-priced evening in a faded setting. Ranelagh Gardens were long gone. By the 1830s even the Vauxhall Gardens were looking their age, although an eighty-foot-high 'Cosmorama' of the Bay of Naples, with Vesuvius erupting, freshened up the programme for a while. In fact, the original Cosmorama was a kind of peepshow, a room with paintings let into the walls, displayed behind convex glasses.

Edward Cross's venture, the Surrey Zoological Gardens in Walworth, was on the south bank of the Thames. Originally a well-stocked and extensive menagerie, aiming to compete with the Regent's Park Zoological Society in particular, Mr Cross soon diversified these entertainments to take on Vauxhall. The setting, with attractive gardens and a three-acre lake, was ideal for a programme of dramatic public attractions. Sometimes Cross's ambitions ran away with him. In 1838, after hours of waiting for a promised hot air balloon, the enormous crowd saw a boat appear bearing a banner with a message, 'The Balloon cannot ascend but to compensate for the unavoidable disappointment an eruption will take place at dusk'. The enraged ticket holders fell upon the grounded dirigible, hurling anything they could find at the inflated envelope and cutting its guy ropes. But what they had been offered – and on this occasion refused – became a well-liked staple of the programme. The 'eruption-mongers', as Richard

Altick so neatly christened them, rotated the famous volcanoes, Vesuvius, Etna and Mount Hekla, around the amusement parks. And, with immaculate timing, the volcanoes alternated too. When Etna erupted in 1852, Vulcan and the Cyclops shot back into vogue on the playbills for London's halls and gardens of entertainment.

Cross had introduced the 'modelled panorama', a life-size, three-dimensional setting for the performance in question, whether it be an eruption or an illuminated scene such as the Fire of London or St Peter's Rome on a feast day. No longer simply a manipulated image, the three-dimensional panorama enfolded the audience so that the scene became far more convincing. Vauxhall followed suit. The genre had a considerable advantage over its predecessors for it could be admired by day (allowing for valuable extra ticket sales), during which the immense painted canvas was reflected in the Bay of Naples, the Thames or the Tiber – depending on the subject – before becoming the backdrop for dramatic pyrotechnics by night. Modelled panoramas countered Charles Dickens' criticism of 'pictorial entertainment on rollers' but he had other criticisms of Surrey Gardens.

In 1838, when Dickens was editing *Bentley's Miscellany*, an anonymous poem appeared. 'Good Mr. Cross! we hate the fuss/And flames of your Vesuvius' it began, admonishing him for the explosive level of noise which disturbed sick children, drove caged animals to desperation and made sleep impossible. It ended: 'To be good-natured, and your name belie/ Indulge no more these furious fiery fits/Let such freaks cease/ Blow up your Mount Vesuvius—all to *bits*/And prithee let us have—"a Little Peace!"' Dickens, clearly the author, did not prevail. Volcanoes and fireworks were too popular to drop from the programme, however much they might madden the wretched caged animals and leave the local population insomniac. After Cross's retirement, the Vesuvius show was smartened up to capitalise on the news of yet another eruption on the Bay of Naples.

While Vesuvius was being created, at many scales and in many locations, the actual volcano drew a growing public – attracted by what they had seen and read about it. In the early spring of 1845, Charles Dickens finally saw the (currently dormant) volcano for himself, climbing it in a party of six, including his wife Catherine and daughter Georgina. He already knew the scenario by heart for he had read his friend Lady Blessington's account and well knew that the journey up Vesuvius was little more than a stage peopled by endearing caricature figures. His own Italian literary sketches soon followed, published weekly in the *Daily News* and then as a book, *Pictures from Italy*. He leant heavily on Lady Blessington's travelogue, *An Idler in Italy*, published in the 1830s but in reality a heavily embroidered account of her visit ten years before. Those who thirstily gulped down the Anglo-Irish writer's witticisms were not looking for practical tips on timetables, currency or how to avoid thieves. Her entertaining potboilers were pumped out to keep the household solvent but she did not have Dickens' added bonus – vignette woodcuts by Samuel Palmer whose experience of Vesuvius was coloured by being trapped with his wife (an enthusiast for all things volcanic, as he discovered) in a cottage just outside Pompeii during the eruption of 1838.

Sometimes Vesuvius spun its own myths; readers of a London gazette in the 1830s were informed that Naples was lit by gas supplied from the depths of the volcano. Yet

Marguerite Blessington's own guide had been an authoritative and influential figure, the antiquarian Sir William Gell. His *Pompeiana* was first published in 1817–19 and ran into many editions. The illustrated volumes provided crucial background for Edward Bulwer Lytton's novel, as well as for John Martin's immense painted cataclysm showing *The destruction of Pompeii and Herculaneum* (1821). Lytton acknowledged his debt to Gell by dedicating the *Last Days of Pompeii* to him.

Lady Blessington's ascent of Vesuvius began, as everyone else's did, at the house of Salvador, 'the most esteemed of all the guides', where a predictably Hogarthian scene was taking place. Their party of eight had attracted an escort of sixteen donkeys and twice as many putative guides, together with their 'mothers, wives, sisters, daughters and aunts'. This gaggle, the women gaudily dressed in picturesque tattered outfits, set about touting for business. Salvador doggedly chose eight animals, to be subjected to a torrent of verbal and physical abuse from the owners of those rejected. By the end everyone was falling about with laughter and their party set off happily enough. Reaching the Hermitage far above, just as the bells were ringing for prayers, they enjoyed the hospitality (biscuits, Lachryma Christi and couches) provided by two sociable old monks. Parting from Gell (who was disabled by rheumatoid arthritis) they continued on up in the usual way. We can well believe that Lady Blessington, like her equally plucky compatriot Katherine Wilmot, waved away the kitchen chair converted into a shaky sedan that was a lady's prerogative preferring to be hauled up the pebbly mountainside on foot, alongside her male companions.

Lady Blessington added a footnote to her account, a description of the liquefaction of the blood of San Gennaro.

The events of 19 September (as those of May and December) were, she considered, although she was a Catholic, an 'extraordinary example of superstition', untouched by the education which 'has so much dispersed the mists of error and ignorance'. She took her readers through the process. Three keys secure the precious relic, each entrusted to different 'bodies of the state'. The blood, looking much like a 'morsel of glue', is brought into view and the officiating priest prays with fervour, 'with exclamations interrupted by his tears and sighs'. A large candle is placed on the altar and the priest moves the phials (in a single reliquary) close to the heat. In the chapel the descendants of the saint (she estimates about a hundred women) are shrieking and 'abused the saint in the most approbrious terms, calling him every insulting name that rage or hatred could dictate'. She and her party approach the altar, only to be set upon by San Gennaro's 'unnatural descendants'. Abused as heretics who might prevent liquefaction, they are forced to kneel and remove their bonnets. Within minutes the blood begins to liquefy and the congregation erupts, weeping and shouting with joy. Even a group of phlegmatic Austrian soldiers seem impressed while a gaggle of Chinese boys in training for the priesthood smile and weep; learning superstition rather than 'the pure principles of Christianity', as she puts it.

Lady Blessington remained doggedly convinced that liquefaction occurred either through the warmth of the hand or of the large candle. 'I confess I left the spot an unbeliever of the asserted miracle.' But she was, of course, playing to an often rabidly anti-Catholic readership. Her account was as much of a construct as a cork model of Vesuvius in a curtained room in central London. Mark Twain had even less patience

18. The Hermitage, the favourite resting place for visitors, shown in a lithograph of 1831 by John Auldjo (who portrays himself in the right-hand corner). Going up or coming down, travellers could pause at the Hermitage for a glass of the famous Lacrima Christi or a comfortable chair on which to rest. Presiding over the establishment was a 'hermit', a gowned figure (sometimes more than one) who appeared to be in holy orders but was, in practice, little more than an avuncular innkeeper.

with 'one of the wretchedest of all the religious impostures one can find in Italy'. In his view, the entire episode was a sly way of filling church coffers at regular intervals.

Anna Jameson, another Anglo-Irish writer, disguised her own account of an ascent of Vesuvius in 1821 as thin fiction in a novel entitled *Diary of an Ennuyée*. She peppers an exceptionally acute travel journal with frequent asides to remind us (and herself?) that she is describing the romantic travails of her heroine, a tragic, Corinne-like figure. Yet despite her attempts at artifice, Jameson has a sharp, fresh eye. She watches a distant line of travellers creeping along the rim of a stream of lava as they ascend the volcano, the six or eight lamps, 'contrasted with the red blaze which rose behind, and the gigantic black back-ground, looking like a procession of glow-worms'. She describes the immense wild aloes that obstruct the path on the way down, their monstrous fleshy limbs illuminated by the red glare off the lava. And, she watches how the guides wave their torches about to stop them from being extinguished, lighting up the dark eyes gleaming out of their striking faces, such a contrast with the 'tall martial figures' of the Austrian military as they stand, sit or lie on their cloaks on the mountainside, watching the eruption 'in various attitudes of amazement and admiration'. After staring into the flame and glare of the five flaring lava torrents that were creeping down the mountain, she needs to rest her eyes by looking into the shadows, the blackness of the landscape everywhere speckled with the lights of the crowds who were 'up and watching that night'.

It is Anna Jameson, too, who fleshes out Salvatore, the chief guide. He had been on the mountain for thirty-three years and tells her that he still corresponds regularly with Sir

Humphry Davy. She provides a rare and humane glimpse of one of those individuals who, rarely given credit for their knowledge and expertise in volcanic matters, were the unofficial guardians of Vesuvius. Widely vilified as money-grubbing tricksters, some of the guides deserved considerable credit for their skills and experience. The lava flows were particularly fierce at the time of Jameson's visit, but she was determined to press on, necessitating Salvatore's close attention, as the most expert and experienced guide.

He showed her how to avoid an immense broiling rock as it hurtled towards them – by staying still. His intelligent conversation and his care and solicitude for her were winning. She makes the guide, so carelessly caricatured by others, a highly knowledgeable and professional man. He identified some of the differences between Vesuvius and Etna (which he had visited during an eruption), judging Vesuvius 'a mere bonfire in comparison'. Yet he and his fellow guides had enormous pride in the 'performances' of their volcano, much 'as the keeper of a menagerie is of the tricks of his dancing bear'. They drew her attention to every new burst of fire or fresh stream of lava, shouting, 'O veda, Signora! O bella! O stupenda!'

⧗

Dickens (no doubt with Bulwer Lytton in mind and aware of how well-trodden the ground now was, in print) shifts the focus somewhat. He sets his Vesuvius against a backdrop of Pompeii and Herculaneum from where 'the genius of the scene' can be glimpsed between the columns, looming over fallen walls or far away beyond the vanishing point of long residual streets. The volcano, 'the doom and destiny of all this beautiful country, biding its terrible time', is crowned with snow and there's a full moon in prospect.

Dickens' plan is to enjoy Vesuvius with 'sunset half-way up, moonlight at the top, and midnight to come down in!' and by four in the afternoon we find him in the stable yard of Signore Salvatore (by the 1840s surely another guide of the same name?) whose gold-banded cap signals his rank as captain of a small army of mercenaries. He finds around thirty guides argumentatively preparing six ponies, three litters and a variety of 'stout staves' for them. When Mark Twain prepared to ascend the volcano thirty years later, the bargaining took an hour and a half, after which he was led by 'a vagrant at each mule's tail who pretended to be driving the brute along, but was really holding on and getting himself dragged up instead ... I began to get dissatisfied at the idea of paying my minion five francs to hold my mule back by the tail and keep him from going up the hill, and so I discharged him.' His journey proceeded much quicker thereafter.

Yet for all his picaresque adventures on the foothills, Dickens, like many others, found himself surprised by his visceral, shocked reaction to the actual landscape of Vesuvius, so unlike anything even such a worldly visitor had experienced before. As they ascended, he was faced by a bleak, bare plain with random lumps of rust-coloured lava lying as if suddenly abandoned, 'as if the earth had been ploughed up by burning thunderbolts'. For Twain, faced with the same stretch of tormented ground, the old lava flow was 'a black ocean which was tumbled into a thousand fantastic shapes – a wild chaos of ruin, desolation, and barrenness ... all this stormy, far-stretching waste of blackness, with its thrilling suggestiveness

19. John Auldjo's 1831 lithograph shows a dapper visitor, dressed rather for town than mountain, with a handkerchief pressed to his nose against the strong whiffs of sulphurous gas that are coming, along with much debris, from vents on what appears to be a newly sprung fumarole below the volcano.

of life, of action, of boiling, surging, furious motion, was petri-
fied!' When Dickens arrives, the sun is going down and the
cone looms, almost on the perpendicular, above. Mark Twain
guessed the height as some thousand feet, almost too steep to
climb and certainly not by a man riding a mule.

Dickens' ascent is helped by reflections bouncing off the
snow, luckily since the guides have dispensed with torches
ahead of the moonrise. (Twain went up at dawn). The litters
are allocated; two for the women in Dickens' party, one for
a 'rather heavy' gentleman from Naples. He alone, apparently, requires a lifting party of fifteen, the ladies six each.
Fellow travellers, for the purposes of a good romp on the
page, tended to be either effete or overweight, sometimes
both. In this genial *commedia dell'arte* all the European ste-
reotypes play their roles to perfection. Ahead of the walkers,
the litters tilt this way and that, their porters slipping and
stumbling with their charges – at one moment the large Neapolitan appears to be virtually upside down. Then the moon
rises and with it the spirits of the party. Brilliant moonlight
illuminates the mountain and everything below – the sea, the
city and its rim of vulnerable villages. By the time they reach
the upper level, the rim of Monte Somma, they can see hot
sulphurous smoke billowing out of crevices. Not far away the
active crater is flaming, spitting stones and cinders, which fly
out like feathers and drop down like lead. 'What words can
paint the gloom and grandeur of this scene?' All is confusion;
smoke blots out the moonlight, the sulphur is suffocating
and the ground underfoot is hazardous and full of rents, the
guides are shouting and the mountain is roaring. Irresistibly
drawn to the vortex, Dickens and a companion now crawl
to the crater's edge on their hands and knees, with Salvatore

beside them. The other guides warn of their danger, frightening the rest of the party. But despite the terror of toppling into the vat of flame below, they stay put, craning into the 'Hell of boiling fire below' before rolling away, scorched and blackened, their clothes alight in places.

The descent of Vesuvius is always played in a different key. By sliding into the ash, the walker usually had an accumulated break-fall and could dash down in no time at all. But on this occasion the ash is covered by ice. The guides link hands and form a human chain, and the terrifying downward journey begins. The women get onto their feet, sandwiched between a pair of guides, Dickens' 'Mr Pickle of Portici' is offered a similar arrangement. He refuses and continues with his fifteen bearers, on the principle that they are unlikely to fall at once. The procession is gingerly proceeding down when the fat Neapolitan slips and rolls all the way downhill – reminding Dickens of a cannonball. Then two of the guides slip too. Later it emerges that all have survived, though bruised and dazed by their terrifying plunges. On their return Dickens' group finds a warm welcome; a French party has fared less well and one man has broken a limb.

Everyone seeking wide readership made much of Vesuvius – with cardboard characters, colourful prose and theatrical overstatement. They were evocative prompts for those who would never see Vesuvius for themselves. As entertainments at home became ever more extravagant, writers were forced to follow their example. James Holman, the so-called 'Blind Traveller', climbed Vesuvius in June 1821 while it was active, making much of his determination to 'feel' what others could see, although as he put it, 'I see things better with my feet'. Salvatore told him that he would let the King of Naples know that he had guided the first blind man to the crater and also encouraged Holman to plunge his cane into the hot lava, which melted the ferrule. He kept it as a memento. The shifting emphasis on travellers with a difference – be they intrepid women or men with disabilities (Holman claimed to have been accompanied by a deaf man) – was rather like reality television, always on the look out for a new angle or extreme.

⧗

Such accounts played to appetites for contrived excitement but also helped stir popular tourism in the age of railways and steamboats. Albert Smith's 'Ascent of Mont Blanc' at the Egyptian Hall had reportedly given a decided fillip to English visitor numbers at Chamonix. Eventually Smith's charismatic performance was replaced in 1857 by an evergreen if more static staple, an evocation of the Bay of Naples and Vesuvius. By now it was difficult to hold an audience without special effects. In the late 1880s the housing reformer Octavia Hill despaired of attracting her Southwark tenants to evening lectures 'of the more interesting kind' since she could not offer them 'magic lanterns or explosions'.

Explosions were a lucrative business for Pains, a family firework manufacturer who learned their trade from eighteenth-century European firework masters to royalty and then built up their enterprise in the amusement parks of London in the 1840s, later expanding to the USA and beyond in the 1870s. Although their repertoire was wide, the most popular of all their pyrodramas, as Nick Yablon terms them, was loosely based on *The Last Days of Pompeii*. This helped to re-establish volcanic catastrophe as a theme for outdoor display which then developed its own momentum. In 1893 the seaside resort of West Brighton, easily accessible from New York, was put at risk by a furious fire that had begun in Pain's Amphitheatre. The reflecting pool (such a central part of the scene, being the Bay of Naples) saved the day, serving as a firebreak and providing invaluable water for the fire engines. While the grandstand (with seating for 10,000) and the scenery for 'Nero' were lost, all reportedly uninsured, the stores of explosives, costumes and staging for the Vesuvius show had all been safely carried over to the other side of the water. Soon after, Pains were back in business. With little sense of irony, they continued to evoke the serious conflagrations of history – including the Great Fire of London, the Burning of Moscow and the Eruption of Mount Vesuvius. Nightly, actors and musicians costumed according to the period and event in question, appeared in front of elaborate scenery. Then the entire panorama would be consumed, seemingly by flame, to culminate in a spectacular firework display. These resplendent entertainments were choreographed nightly to pass the summer evenings for the holidaying city people who filled the immense seaside hotels. James Pain took versions of the show to more than thirty cities around the USA; and then in the winter season it all decamped to Australia, where audiences in Melbourne watched these epics in adapted cricket grounds and pleasure gardens.

In 1904 the 'Destruction of Pompeii' was presented at Coney Island's Dreamland amusement park with new electrical effects of extraordinary complexity and novelty, costing hundreds of thousands of dollars; while at a spectacular firework display mounted by Pains at London's Alexandra Palace in the summer of 1906, 'the Eruption of Mount Vesuvius'

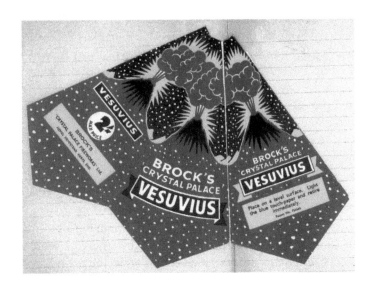

20. Brocks' Crystal Palace Vesuvius, well designed, brightly coloured packaging for a themed firework, costing two shillings. Brocks, a London firm, had been in business since the early eighteenth century and volcano themed fireworks were always popular. An earlier version, costing one penny, was described as a 'grand realistic eruption'.

reigned supreme – for the volcano was active again. A century later, the BBC paid its own homage to Bulwer Lytton in a memorable episode of *Doctor Who*, 'The Fires of Pompeii' in which the Doctor and his assistant are caught up in the destruction of Pompeii. The dramatic events were filmed in 2007 at Cinecittà Studios in Rome on an existing film set depicting ancient Rome. Enormous quantities of ground-up cork were used to simulate the ash cloud raining down on the actors.

Even modern generations used to the effects of sophisticated cinematic animation have not entirely lost their taste for outdoor displays, as long as the presentation meets their heightened expectations. The crowds viewing the volcanic attraction on the forecourt of the Mirage Hotel in Las Vegas in the 1990s had complained of the unpleasant smell of the natural gas used in the explosions. But when The Volcano reopened in December 2008, with nightly shows between 6 p.m. and 11 p.m. (every hour on the hour), that problem had been neatly solved. Now the scent of pina colada wafts over the crowd. Other changes have been made too. It has become a 'primal volcanic environment of sound, light, music and heat' and whereas in the outdated 1989 version the volcanic crater had been the central attraction, now 'flame-expression technology' shoots fireballs high into the air, the lava flows and, finally, the eruption spreads to ignite the surrounding lagoon where, in the technical director's words, 'spectators will feel the heat on their faces'. The sound track, without a whisper of irony, is provided by the drummer of the Grateful Dead.

Long before the advent of television, let alone the creation of Las Vegas, one wintery day in a small town in a drab part of industrialised north-east France, two schoolboys kneel, engrossed. They are bending over a toy theatre, its auditorium made out of a wooden crate lined with packaging paper. Intently, they prepare the *pièce de résistance* of their repertoire, the eruption of Vesuvius, for an eager domestic audience. It's an apt subject in 1888, since news of yet another eruption of the volcano has recently spread around the world. Back in Bohain-en-Vermandois their staging is realistic, convincing the nose as well as the eye and ear of the audience by the suitable addition of sulphur and saltpetre, donated by a trusting father whose work allows him access to such things. Those ingredients, sufficient to make gunpowder but for the nitrate, will give the explosion a life of its own. The Bay of Naples, for these purposes, is evoked by brilliant aquamarine painted paper. The waves meet the shore with foaming white brushwork. As a backdrop, the lit Vesuvius belches impressive quantities of smoke and flame. A violinist plays, intensifying the drama.

The maker and designer of the stage set is young Henri Matisse whose lifelong best friend Leon Vassaux is playing the violin. As an old man, Vassaux remembered their homemade entertainments and wrote to Matisse to remind him about them. The light, the colour and the heat, the evanescent magic of the volcano and its glorious southern setting were to be Matisse's first taste of the Mediterranean.

6

....

REAL GEOLOGY, NEW FOCUS

In 1802, the great engineer James Watt's young son Gregory, racked with consumption, climbed Vesuvius. His colleague and supportive companion in this risky venture was William Maclure, later considered the founding father of American geology. Despite a turbulent youth, Gregory Watt had already shown great vision, having persuaded his recalcitrant father to invest in the novel gas lighting system he had seen in Paris.

In a fascinating brief episode which was to bring the worlds of romantic literature, radicalism and innovatory science together, Gregory Watt had also recommended the very young and untried Humphry Davy for his post as Superintendant at Thomas Beddoes' Pneumatic Medical Institution in Bristol – where he met the Romantic poets Coleridge and Southey. Their experiments with gases were, albeit inadvertent, steps towards the development of anaesthesia. Aged nineteen, Davy was inspired to write an ode about his experiences, the 'inward transports' of inhaling nitrous oxide (laughing gas).

Watt's own fame was posthumous, for as Davy wrote, devastated by his patron's early death, 'he ought not to have died' for he 'would have been a great man'. Yet in his last two years of life, on his return from Naples, he worked indefatigably, always inspired by his remembered views of the crater. He was busily working on a geological map of Italy, but Watt's only publication, on the cooling of basalt, was to be enormously influential. Based on his experiments with melting and freezing rock, his observations would finally help to decide the long battle which had been raging between those geologists known as Neptunists, who ascribed the creation of the Earth's crust to the action of the seas, and those known as Plutonists, who considered that basalt, an igneous rock resulting from crystallising magma drawn from the very heart of the Earth, held the key. Acceptance of the latter view presented a stern challenge to biblical chronology and the story of the Creation.

In 1819 and then again in 1822 there were eruptions of Vesuvius. The emissions were less intense than on some previous occasions and the dangers, though not to be underestimated, foreseeable. Usefully enough, from the point of view of a new generation of European scientists, it was returning to a cycle of relatively continuous activity, much as in the days of Sir William Hamilton. Like him, the volcano-watchers, now scholars and professionals rather than enthusiastic amateurs, would be almost spoiled for opportunities to observe an active, if not an erupting, volcano. Geologists had had their own learned society since 1807 and there was now a Chair in the subject at Cambridge. More followed. In parallel came growing academic recognition for closely related topics such as mineralogy and palaeontology. Vesuvius, long seen as a giant chemistry set, was now all set to become a crucial research laboratory where geologists and their colleagues could do fieldwork with essential evidence to hand. So accessible for all Europeans, the enterprise also proved gratifyingly and fruitfully collaborative.

The two men who published the leading works on volcanic activity in English of the first half of the nineteenth century were exact contemporaries. Most of the first-hand evidence in their revelatory accounts came from either the extinct volcanoes of land-locked central France or the obligingly active volcanic regions on and near the mainland around Naples and Sicily. Together they provided a sequential picture, linking ancient events and current activity. The evidence of stratification began to point to the truth about the formation of the Earth, giving geological history a timescale that beggared belief and dented scriptural certainties.

In the winter of 1817–18, twenty-year-old George Poulett Thomson stayed in Naples with his family. What he saw there transformed his ambitions and interests, for he found himself in the very spot that marked, in Martin Rudwick's words, 'the canonical intersection of geology with archaeology, geohistory with human history'. After graduating, he embarked on two years of European fieldwork, returning to Vesuvius just in time for the 1822 eruption, now called George Poulett Scrope, having taken his wife's name on marriage in order to inherit her comfortable fortune.

Building on theories already advanced by others, but either overlooked or challenged by those with different (and generally prevailing) opinions, he finally established the key role of volcanoes in forming the earth. Lava and ash were driven to the surface by the pressure of intense subterranean heat (from still unidentified reserves of magma at great depth), subsequently cooled and readjusted by the seismic movement of tectonic plates, a factor leading to earthquakes and itself not fully understood for more than another century. Tell-tale signs of marine life high on the columns of the Temple of

Serapis at Pozzuoli, in the volatile Phlegraean Fields, told of changing sea levels and pointed to movement in the earth's crust, both up and down, in relatively recent times.

These processes, aided by water and steam, had been the agents of the earth's continuously and necessarily changing form over time, together with erosion and overlay of sediment. Volcanoes as such could be seen as a 'safety valve to the globe' – a conclusion not so different from Kircher's graphic notion of two hundred years earlier – or for that matter Empedocles' from some millennia before. With Scrope's *Considerations on Volcanos* (1825) and a subsequent volume on the geological formations of central France, especially those all-important dead volcanoes of the Auvergne (long ago identified as crucial evidence by Nicolas Desmarest), the lingering mystery of the origins of volcanic basalt was solved – and thus began the testing affair of equating these conclusions about the forma-tion of the universe within a timetable set by the Creation myths. For the time being, the Neptunists' (water) theory had been decisively toppled by the Plutonists' (heat) argument. Ironically, modern geologists now suggest that many sedi-mentary rocks have been laid down under water; the outcome was never likely to be quite that black or white.

Scrope's exact contemporary and friend Charles Lyell was a classicist who was drawn to geology while up at Oxford, largely because of the remarkable and idiosyncratic lectures of William Buckland, a cleric who held a readership in geology and was therefore at pains to equate his scientific observations with his theology. For Lyell, familiarity with the object under study rather than speculation was his credo (just as Hamilton's had been). In 1824 he read a paper at the Geological Society setting out his lifelong adherence to the crucial importance of 'true causes' in geology; what could be evidenced rather than speculated upon. He began his field studies in southern England and then continued them in the Paris region, while supporting himself as an essayist and barrister.

In 1828, after a prolonged fieldwork trip in the Auvergne with fellow geologist Roderick Murchison and his wife, Lyell went on alone to southern Italy. Etna was his objective. But while his passage to Palermo was delayed, he filled the time by studying what he could easily reach around Naples. He climbed Monte Somma, examined Monte Nuova and the Phlegraean Fields, explored the archaeological sites of the lost cities and, finally, ascended Vesuvius which helpfully laid on an eruption, just as it had for Scrope in 1822. But the violence and unpredictability of the 'constant ejections' prevented him from getting anywhere near the crater. Also nearby was Craufurd Tait Ramage, who was tutor to the children of the British consul. Having taken one of his small charges to safety, he ventured back and just as he did so, a section of the crater edge opposite fell away, offering a graphic illustration of his own danger, balancing on the rim of a collapsing volcanic cone.

For Lyell, everything he observed on this journey was fitting into a pattern, a puzzle that was finally beginning to make sense. As he told his father, what was once a strange script was now a language that he could readily interpret. It was an age of decipherment in every field – from archaeology and paleontology to geology. Back in London he embarked on a major project, his three-volume *Principles of Geology*, to be published from 1830 to 1833. One reviewer remarked on the way two divergent interests seemed to meet in the younger generation, educated as the classics met the sciences, busily 'geologizing with a Virgil in one hand and a hammer in the other'.

The hammer-wielders were now in the ascendant over the classicists, and working inexorably towards their theologically uncomfortable conclusions, in which the world's origins were envisaged on a timetable of some millions of years, rather than the comfortable 6,500 year biblical history, beginning (geologically speaking) with the Great Flood. Genesis, Noah and the geologists were on a collision course. Convulsions below ground were mirrored above. In 1831 Lyell was awarded the chair in geology at King's College, London, a new and avowedly Church of England institution; Lyell's tenure there was to be less than three years and his lectures at the Royal Institution lasted just a year. The tensions, largely unstated, were building and though Lyell had said privately that he wished 'to free the science from Moses' he, in common with many of his contemporaries, had great difficulty coming to terms with the logical conclusions of his own observations.

Nevertheless, the subtitle of the first edition of the *Principles* was 'an attempt to explain the former changes of the earth's surface, by reference to causes now in operation', underlining Lyell's insistence that geology must take note of modern events, but only as part of a continuum of change. Fossil remains, often exotic and dislocated, were visible evidence of events and phases in the earth's development that might otherwise appear incredible. By the time that Queen Victoria ascended the throne, Lyell was President of the Geological Society, an established natural scientist of the Regency period whose correspondents (he was impressively multi-lingual, facilitating the cooperative nature of scientific endeavour) were the French, German and Italian leaders in the field. *Principles* was widely translated while his more general introduction to geology, *Elements of Geology* (1838), succeeded in bringing the subject to a less specialist readership.

Lyell continually revised his work over some forty years, while Scrope effectively vacated his place in the field for his own brand of liberal politics, not returning to the rapidly developing subject until the mid 1850s and then only as a prolific observer and reviewer. As the title made clear, Lyell's *Principles* (which reached its posthumous 12th edition in 1875) shared the broad ambitions of Isaac Newton's *Principia Mathematica* albeit in a different field. The essential argument was that, seen within a geological timescale, the episodes and dramas of our planet were not sporadic catastrophes but a continuous process, much of which was unrecorded, obliterated or overwritten by later events. As Stephen Jay Gould puts it, from Pliny onwards the idea had taken root that the earth was 'ruled by sudden cataclysms that rupture episodes of quiescence and mark the dawn of a new order.' Lyell's so-called 'uniformitarian' argument, as already advanced years before by the ground-breaking Scottish geologist, James Hutton, was precisely the opposite to that of the 'catastrophists'. It depended on an extended timescale, with the resulting challenge to the theological view of creation.

Thus the physiognomy of the area around Vesuvius could be seen as the result of relatively gradual change, as various cones had risen, altered their form or deferred to new landforms, rather than springing from the ground simultaneously, in Lyell's happy phrase, 'like the soldiers of Cadmus when he sowed the dragon teeth.' With a certain lofty (and arguably irrelevant) superiority, Lyell also pointed out that the citizens of Naples and the wider Campania had always been more

21. Lithograph of Charles Lyell as a young man, around the time he first visited Vesuvius. Prompted by his experience there, Lyell became the pre-eminent geologist of his generation.

22. James Nasmyth's probing intelligence emerges in this engraved portrait, taken from his autobiography. Ironmaster, inventor and moon-gazer, Nasmyth considered Vesuvius the home of Vulcan, god of his own craft.

severely and continually threatened by moral than physical disaster – 'disastrous events over which man might have exercised a control, rather than … inevitable catastrophes which result from subterranean agency.'

Lyell argued that undue credence was given to surviving evidence and thus science was relying on an imperfect record. He drew his illustration from the history of Vesuvius. If, by chance, another volcanic layer had been superimposed on the ruins of Herculaneum, all the evidence would be obliterated, then overlooked in the absence of any physical clues. (In 1906 Vesuvius immolated, for the second time, a particularly fine 'suburban' villa in the hills beyond Pompeii; the only surviving record is an excavation undertaken between 1903 and 1905.) Lyell's point had the deepest of implications, yet he only gave Charles Darwin's *Origin of Species* (1859) his 'qualified approval' in the tenth edition of the *Principles*, almost a decade later. His reticence was a considerable disappointment to Darwin since he had put Lyell's 'noble views on "the modern changes of the earth, as illustrative of geology"' very much centre stage. Darwin closed his chapter on the 'Imperfection of Geological Record' hinting at Lyell's difficulties, for, he wrote: 'I look at the natural geological record, as a history of the world as imperfectly kept, and written in a changing dialect; of this history we possess the last volume alone … Of this volume, only here and there a short chapter has been preserved; and of each page, only here and there a few lines.'

Geologists were not the only ones to be intrigued by the elemental forces at work, generally unseen but occasionally glimpsed at the very core of Vesuvius. James Nasmyth, the great iron master and inventor, climbed the volcano in 1865. He stood well to the windward side when he reached the edge of the crater ('thus out of harm's way') but, though wary, he was drawn by the optical effects, vivid colours which he knew by their chemical parentage to be 'the result of the sublimation and condensation on their surfaces of the combinations of sulphur and chloride of iron, quite as bright as if they had been painted with bright red, chrome, and all the most brilliant tints.' Around this diabolical interior swirled vapours which emanated from every fissure and crack within, while far below boiled the red hot lava. Nasmyth wanted to test the depth, being a bit of an overgrown schoolboy as well as an esteemed engineer. He heaved a great lump of lava to the edge and rolled it over but to no effect. The depth was either too great or molten lava had broken its fall – nothing could be heard. In a gracious gesture, the brilliant designer of the steam hammer and patent holder for hundreds of ingenious machine tools, tied a card from his Bridgewater Foundry to a piece of lava and hurled it down 'as a token of respectful civility to Vulcan, the head of our craft.' In all, he spent five hours entranced on the edge of the crater, writing and drawing what he saw.

Soon after this visit Nasmyth retired from business in order to pursue his other interests, in particular the close study of the moon. In his book on the subject, he expanded on his observations of Vesuvius. As he wrote in his autobiography, volcanic action 'has been, as it were, the universal plough! It has given us picturesque scenery, gorges, precipices, water-falls. The upheaving agent has displayed the mineral treasures of the earth, and enabled man, by intelligent industry, to use them as mines of material blessings. This is indeed a great and sublime subject.' For James Nasmyth, as for the ancients, Vesuvius was the ultimate foundry and a site potentially holding the secrets of the universe.

By the time Nasmyth visited, L'Osservatorio Vesuviano had been established for some years and the mountain subjected to new levels of scientific scrutiny. Such an observatory, located high on the mountain itself, at a location called Eremo, had been proposed by the Bourbons back in the early 1800s and support offered by the prestigious Royal Academy of Sciences in Naples. Teodoro Monticelli, the chemist for whom Dahl had painted his dramatic eruption in 1820, was an active vulcanologist who played a key part in the early discussions but, on being accused of consorting with revolutionaries, he was forced to move to Rome. A sizeable part of his mineral collection had already reached the British Museum by 1823, with Humphry Davy as mediator, and was bought for more than £500.

Eventually King Ferdinand II gave the observatory project the go-ahead in 1839 and building began in 1841. The first director was the distinguished physicist Macedonio Melloni, who was appointed at the instigation of the internationally respected German geologist and geographer Alexander von Humboldt. A radical who was sent into exile for some years, Melloni received the Rumford medal of the Royal Society while abroad in 1834 and was made a member of the French Académie des Sciences. Yet Melloni's speech at the official opening of the as-yet incomplete Observatory was studiedly diplomatic, referring to the mysteries that still lay beneath their feet and his dedication to lifting 'questo grave velo' (this heavy veil), which had so far hidden the sources of the earth's heat. Observation was to be the key, an institutionalised version of Hamilton's decades of record, and Melloni presented careful neutrality in the face of the evidence. It was neither the time nor the place, within a profoundly Catholic milieu, to proclaim heresy.

The handsome neoclassical structure, with its stone portico and regal flight of steps, as well as a campanile-like outlook tower, was the first dedicated volcanic observatory in the world. Wrapped around it were terraces providing external observation platforms. It stood on the Colle del Salvatore on the site of the old Hermitage from where countless visitors had set out on foot to scale the mountain, well fortified with Lachryma Christi. Building so close to Vesuvius put the Observatory, in the view of many experts, in the likely path of lava flows from an eruption but such proximity was crucial in order that essential readings and research could be easily undertaken.

Hardly had the Observatory opened its doors in early 1848, than the institution fell prey to the continued ructions of the Risorgimento, Italy's long struggle to become a single nation, embracing attempts to reform the Constitution and unify the disparate parts (and many powerful players) on the peninsula. Italy was swept up in the Europe-wide revolutionary tailspin of that year. Six months later Melloni, as a marked radical, was removed from his post by order of the Bourbon court. Meanwhile Pope IX had fled from the Vatican, escaping the imminent Roman Republic and took refuge in the south, at Gaeta, where he remained from November 1848 until moving to Portici almost a year later, living in the royal palace until his return to Rome some months after the dissolution of that short-lived Republic, in April 1850.

Yet Melloni's 'exile' appears to have been relatively nominal. Such was his distinction in the European scientific community that he was sent no further than Portici from where Hamilton had observed Vesuvius to such excellent effect (and where, ironically, he was now a close neighbour of the Pope). Melloni continued his work on radiant energy at the Villa Moretta. No successor was appointed until the year after his death, from cholera, perhaps as a further mark of respect towards the distinguished scientist who was, in all but name and station, still in charge.

Republicanism and science were natural bedfellows and several other distinguished academics had fallen foul of the system. It was not until late 1860, when Garibaldi arrived by train in Naples, that the reactionary Bourbon rulers of the Kingdom of the Two Sicilies finally abdicated and liberal-minded professionals could take up their work again, now without fear of banishment. Yet those who took their opinions on these matters from the Catholic Church still objected actively to many of their conclusions and undermined scientific progress.

Melloni's successor as director of the Observatory was Luigi Palmieri. His tenure began in 1855. In 1859 he established the *Vesuvius Observatory Annals*, which he edited until 1873. Palmieri spent much of his time at the Observatory developing methods of measuring subterranean movement, which he identified as a crucial indicator of any impending disaster. His earliest seismograph, based upon u-shaped tubes filled with mercury that could activate an electrical contact, was for indoor use and was so effective that a version was sold to the Japanese. Later he developed a lightweight, portable version – which was superseded (in time for the eruption of Krakatoa in 1883) by a horizontal pendulum model.

At the height of the 1867 eruption, Palmieri stayed put at

the Observatory for over a week, watching and noting every moment of the volcanic episode. Several of his successors took similar risks; the thrilling culmination of those long years of waiting was not to be lost to petty concerns about personal safety. Meanwhile, at much the same date, Jules Verne's hero in *Journey to the Centre of the Earth* took a short cut only available to a character in science fiction, by entering the core of the planet in Iceland and exiting it at Stromboli.

Observing Vesuvius through a powerful telescope from below in Naples was Mary Somerville, the polymath after whom the Oxford women's college was named. Despite her great age (she was born in 1780) Somerville retained a vivacious interest in 'mathematical and scientific subjects'. Late in life, she and her two daughters had moved from Rome to Naples where she was immediately introduced to the leading scientific figures in the city and elected an honorary member of the Accademia Pontaniana, the ancient learned society of Naples. She was watching the 1867 eruption with close attention.

That November she had seen a new crater appear in the Atrio del Cavallo (the desolate, rubble-filled valley immediately below the existing cone). By the following day the lava stream, which she estimated to be a mile wide and thirty feet deep, was heading for the plain below, setting the vegetation to either side ablaze. Somerville could not risk the climb but remained in station at her window while her daughters went up in her stead. In pondering the causes of this eruption, she wondered if the excessive rainfall of recent months had exacerbated the process, a hypothesis she may have had a chance to discuss with a visitor, the noted Oxford geologist Professor John Phillips.

Much as Sir William Hamilton was to become the first port of call for those interested in the doings of the volcano, so Henry Wreford, contributor to the *Athenaeum* and *Household Words* and *The Times* man in Naples, assumed a similar role for Victorian visitors from home. He gave Phillips an introduction to Luigi Palmieri and recommended his own preferred guide, Giovanni Cozzolini. Phillips arrived in early 1868, in time to catch the later stages of the recent eruption. He published an important account of Vesuvius the following year.

The geologists came and went, always at risk of missing the moment, but Mary Somerville was on the spot. On 26 April 1872, a Friday, she was woken by her daughter Martha at 1 a.m. 'that I might see Vesuvius in splendid eruption'. At breakfast time even she, profoundly deaf as she now was, could hear thunder which, her maid told her, was no storm but the mountain. Quickly, the Somervilles decamped to a hotel with a good vantage point. From there, in the dress circle as it were, they watched an eruption that was quickly becoming, Mary noted, the fiercest in a generation. Another spectator was the traveller Charles Doughty who claimed to have been standing on the summit, alone, when Vesuvius erupted. In his extraordinary, anachronistic prose he recalled the 'vulcanic womb' and the 'belly of the volcano hill' in its 'uncouth travail' – memories apparently summoned up by the contorted sandstone landscape he was describing in *Travels in Arabia Deserta*.

That first day Mary Somerville saw not Doughty's 'pumy writhen slags' overhead, but a scene bleached quite white by the immense quantity of steam billowing from the lip of the crater. She was highly relieved that her daughters had not repeated their excursion to the Atrio del Cavallo since up to fifty people were killed there when a new crater – or fumarole – opened up literally beneath their feet. They were, in Mary Somerville's sober words 'scorched to death by the fiery vapours which eddied from the fearful chasm'. In the city, the railway stations were in chaos as those with the means to flee passed incoming trains crowded with the curious. After a rumour began that an earthquake was imminent, Neapolitans avoided the narrow streets and stayed in the open all night.

The 1872 eruption, the first to be captured on camera, laid waste to the countryside. The deadly ash killed all the vines, fruit trees and crops that the rivers of molten material missed. The human casualties were high. Yet after the destruction, volcanoes always startle by their peculiarly transcendent beauty. One valedictory evening in late May, near sunset, as Mary Somerville was watching 'when all below was in shade, and only a few silvery threads of steam were visible, a column of the most beautiful crimson colour rose from the crater and floated in the air.' She died that November aged ninety-two.

Volcanic activity continued to be a live topic in the fraught arena in which the Victorian gladiators of science and religion met and wrestled. The subject, and sundry opinions, seemed to be open to all, rather like climate change in our own day. The complex arguments are neatly characterised by the naturalist T. H. Huxley in a letter to his friend, the physiologist Michael Foster. As he headed home in April 1872 (ironically

23. This is the first of many photographs to capture Vesuvius erupting and is carefully dated 26 April 1872 and the time given as 3.30 p.m. The photographer is not known but was positioned at a distance across the Bay, from where he could depict the height of the pine-tree shaped, or Plinian, cloud.

enough just missing the actual eruption) he wrote: 'The best sight I have seen in all my travels was Vesuvius ablaze – I went up and had a pocket eruption for my especial benefit. No, one thing in its way was better and that is the Apollo Belvedere – because man with his limited faculties did the latter, and the Almighty to whom it is no trouble the other. But it just occurs to me that the Almighty did both. Confound Pantheism & all metaphysics …'

More fortunate than Foster was Giuseppe De Nittis, one of the Scuola di Resina group of painters. He returned to southern Italy after a stay in France in late 1870 to begin an intense study of Vesuvius, in every light, all weather conditions and, frequently, at very close quarters. He was still there in 1872 to capture the volcano as it erupted that April. But his oil painting of villagers fleeing from the towering ash cloud was, for once, not the result of his own observations on the spot and was probably compiled from photographs. Until that moment, his geological and topographical accuracy, as demonstrated in more than sixty sketches of Vesuvius, had been almost that of the scientist.

Yet at the very moment that geological expertise had become an academic specialisation within those disciplines collectively known as the earth sciences, it was two Anglophone professionals – an Englishman and an American, a doctor and an engineer – who assumed the mantle of the eighteenth-century amateur vulcanologists. Each, working respectively in the late nineteenth and the early twentieth century, was widely recognised to have brought their own particular insight and skill to bear. Expertise built up, much as the lava itself did, changing the very outline of the volcano in every generation.

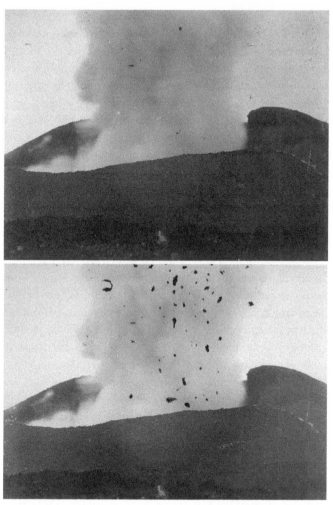

24. The crater of Vesuvius photographed on 6 February 1884 by Dr Henry Johnston-Lavis. He apparently scattered debris onto the negative while it was being exposed, to boost the dramatic effect of the image below.

The Observatory Museum at Vesuvius has a section dedicated to Dr Henry Johnston-Lavis, as great a hero to international vulcanologists as his compatriot Sir William Hamilton, also honoured at the museum. Back in London, a modest room in the Earth Science department of University College also bears Johnston-Lavis's name. Together these two rooms, one in Campania and one in central London, commemorate far more vividly than any carefully lettered monument would, the extraordinary achievements and single-mindedness of the man. His efforts propelled the study of Mount Vesuvius (and with it, of vulcanology) into the twentieth century, leaving the topic better recorded, examined and understood than ever before. In the UCL Rock Room, as his collection is known, there are hundreds of samples of minerals and rocks, in display cabinets and specimen drawers. On the walls (and elsewhere in the university art collection) are some of the gouaches and prints he collected, while his books are held in the main library. His photographs, now conserved and identified as far as possible, show the continuous risks he took, recording violent events with little regard to his personal safety and catching the most ephemeral of landscapes, largely destroyed by subsequent eruptions or weathering, in the most immediate fashion then available to a scientist.

Johnston-Lavis's greatest achievement is displayed on a landing outside the Rock Room: an immense map of Vesuvius and Monte Somma, painstakingly coloured and inscribed, the entire surface patterned with dozens of probing finger-like forms, like spilling pigment. Each represents a flow of lava, eruption after eruption. Some head west, some north-west, some south and others south-east. Every compass point, except the area protected by the jagged

25. Dr Henry Johnston-Lavis in his ceremonial robes, at the time of his appointment as Professor of Vulcanology at the Royal University of Naples, 1892. On the table is a specimen of the mineral that he discovered and named Chloromanganokalite.

spine of Monte Somma to the northeast, is covered with lava flows. Johnston-Lavis's map, based on a military topographic record, was at a scale of 1:10,000. When it was published in 1891, it was an astonishingly accurate record, the fruit of an unprecedentedly long and close examination of the volcano, and bearing the notations of every trace of evidence that he could find upon its surface, to tally with documentary records of earlier eruptions. But within fifteen years, the map was out of date. The eruption in 1906 changed Vesuvius radically, and by 1944 the inner cone of the volcano had been scythed away. Another map at this scale was drawn up in 2003, but this time compiled with state-of-the-art technology, the only modern equivalent to the tireless manual labour of Henry Johnston-Lavis. It will do, for the moment anyway.

Johnston-Lavis was a medical man who, by building up a general practice based around wealthy English-speaking patients who had travelled south for their health and often to mineral spas, managed to combine a respectable career as a doctor in Naples with a spectacular one as a self-taught geologist. From the moment that the Observatory opened, successive directors had been door-stepped by those who claimed intense interest in volcanoes. Johnston-Lavis, who may have seemed no more than another enthusiastic amateur, first introduced himself to Luigi Palmieri in 1879. By 1884 he had read an important paper to the Geological Society of London, the fruit of his extensive fieldwork funded (modestly) by the British Association for the Advancement of Science. In 1892, the year after the publication of his map, he was appointed Professor of Vulcanology at the Royal University of Naples, in recognition of which honour he was photographed in resplendent ermine-tipped robes, standing by a table laden with the attributes of his geological life: a book, a glass-cased mineral (his own discovery to which he gave the unpronounceable name of Chloromanganokalite) and other items.

By 1888 he had amassed all the information he needed to draw up his map. He divided the eruptive patterns into eight phases, that following 1631 being the last. Each section was then coloured (predominantly in variations on red) to indicate the likely age of the lava, and its subsequent development and overlay of soil and vegetation. Almost all lava flows since 1631 are dated. Johnston-Lavis had not had an easy time pursuing his mission. In order to survey the deposits on the mountain, measuring and comparing the strata, a task which took the best part of a year and which he carried out throughout the summer heat, he moved around from one place to another at the foot of the mountain. 'Of actual help I received none, but rather actual obstruction, in some cases of a most regrettable nature, which for the honour of science had better remain undetailed', he wrote, darkly. Perhaps he ran into a wall of professional jealousy: he was, after all, a foreigner and not even a geologist. He was treading heavily on other men's turf. He left Naples soon after for the south of France. In 1914 he was killed in a car crash.

☒

By the time Palmieri died, he had been director of the Observatory for almost fifty years. His successor, Raffaele Matteucci, was only the fourth director in the institution's history. Hardly had he taken on the role when another amateur vulcanologist appeared at the door of the Observatory whose abilities were entirely different, but no less suited to the purpose.

Frank Perret, a prodigious American electrical engineer, had worked briefly with Thomas Edison and then, aged nineteen, set up his own company, called Elektron. In 1887 a motor designed and made by Perret powered the first American electric elevator and his business soon expanded into a large factory in Springfield, Massachusetts. Before long Perret bowed out, turning his attention to the production of motor vehicles before collapsing, aged only thirty-five, with a severe mental breakdown brought on by persistent overwork. He arrived in Naples in December 1903 and paid his first visit to Vesuvius the following month.

The compulsive hold of the enigmatic volcano on an intense, introverted personality is intriguing but not so surprising. In Perret's case, his involvement can be seen as a form of self-administered therapy. As Tom Gidwitz writes, 'he was a gifted inventor, wise in the ways of energy and matter. Vulcanology provided him with a mission. While watching volcanoes for countless hours he could challenge his intellect and indulge his need for solitude, yet play a role on a global stage.' Perret began to ponder means of predicting volcanic activity, even at times of apparent calm. In February 1906, while staying on the volcano – perhaps at the Observatory or in Thomas Cook's nearby hotel – he was disturbed by 'a continuous buzzing sound which seemed to come from below'. Gripping the iron bedstead with his teeth, he found he could now experience the noise more clearly. Perret became convinced that sophisticated microphones were second only to seismographs in their importance as early warning signals of subterranean turbulence, in what would soon be recognised to be an immense magma reservoir.

Frank Perret's other contribution, perhaps more important

26. A record of the 'various aspects of the declining crater-cloud, as seen from the Observatory' photographed by Frank Perret, 14–16 April 1906.

still, was to emphasise the importance of photography as evidence and he carried his folding pocket Kodak camera at all times. His record of the eruption of spring 1906 was in great detail and his expertise second to none. (When Mount Pelée began to erupt again in 1929, Perret hurried to Martinique and stayed to help the traumatised people of St Pierre, fearing a repeat of the devastating eruption and pyroclastic flow of 1902.

Perret settled at Torre del Greco but his eyes were on Vesuvius and he haunted the Observatory, where in 1905 Matteucci appointed him Honorary Assistant Director. His timing was perfect since Vesuvius had entered a new phase of activity that year; Perret had become entirely familiar with its current state but now it appeared that he was about to see a serious eruption. A vapour cloud seen on 4 April was followed by falls of black ash in the city and Neapolitans began to get out their umbrellas. The next day Perret took the railway that now ascended the volcano for much of the distance, as the cloud overhead grew and grew – reminding him, he said, of a giant steam locomotive puffing uphill but continually refuelling. Matteucci and Perret were examining a new lava-vent – a fumarole – low down the mountainside, when they realised that they had to return on foot since the line had been cut.

Back at the Observatory they found the instruments jumping, banging about, as Perret put it, due to the seismic activity. Outside, flashing arcs of light were refracted off the surface of the clouds, while underfoot the ground shuddered continually. The temperature dropped to a ferocious chill, due to the resulting updraught. All of this was, Perret wrote, 'difficult to describe in words befitting a scientific book'. By 3 a.m. on the 8th, the lower station on Thomas Cook & Son's

27. In April 1906 the staff of the Observatory stand at the top of the steps, just cleared of volcanic ash. They had been imprisoned there throughout the eruption.

funicular railway was burning and there was 'an appalling intensity' of electrical activity; the overhead lines of the railway served as conductors to the Observatory. Villages to the north east of the volcano were taking the brunt of the action. Then, before dawn, the cone of Vesuvius began to collapse outwards 'like petals off a flower' – such was the power of the gas pouring out.

Perret returned to Naples later that day to cable his family in the USA. He heard that around 100,000 people had left the city in the last five days. Those that remained, milled around with nothing practical to do but invoke the saints. Having completed his business, Perret felt 'it behooved' him to return to his post at the Observatory. He found his cab was moving against a tide of fleeing people from the villages below the volcano, a crowd seemingly led by children. He soon continued on foot. The scene was, he imagined, just as it must have been at Pompeii. In the near dark brought about by the ash fall, faces peered from doorways, from beneath wagons and any kind of improvised shelter. Soft balls of compacted wet ash – called pisolites – fell constantly. When Perret reached Pugliano he met Signor Mormile, the stationmaster. He volunteered to act as telegrapher at the Observatory and together they followed the railway track back up.

The inhabitants of the Observatory, their numbers now augmented by six carabinieri and their officer, were held hostage by the volcano for more than a fortnight – to begin with unable to even step outdoors. Soon a series of hot, silent, avalanches began. The falls of ash (which, disconcertingly at one point contained caterpillars) had been unpleasant but the new phase, with its attending gas, was more terrifying. On 10 April, the 'prisoners' made their first attempt to go out, and discovered the strange electrical reactions that they provoked, particularly off the metal details on the smart carabinieri uniforms. The following day was pitch dark. Ash fell without stopping, and further shocks could be felt. A unit of engineers came up from Resina to shovel the load off the roof before it brought the entire building down – they shifted six tons of ash. Later that day, when the sun had finally broken through, the Observatory 'hostages' were photographed standing on the porch, the mountains of ash to either side of the steps resembling innocuous piles of snow, an impression intensified by their cheerful tanned faces, like a group of winter sports enthusiasts outside their chalet.

It was, Perret admitted, a tedious period but they were catching an unprecedented view of the action. In addition, that they were known to be surviving upon the very brink of the erupting volcano was a masterly public relations stroke; posters were put up in Naples and neighbouring towns with reassuring and up-to-date news from the nerve centre up on Vesuvius. In fact, the information was of no use whatsoever, for any change in the mountain's behaviour or the weather conditions altered the situation entirely. For example, on 13 April, the *New York Times* reported, the wind changed direction to blow the ash back over Naples; despite the season, the reporter described people rugged up in car-coats, caps and goggles, nervously pacing the streets in gas masks and sheltering beneath umbrellas. Meanwhile, the looters had moved in and the birds had flown out. One doughty German minor royal, the Princess of Schleswig-Holstein, set out in her car on a round of hospital visits only to be forced to walk home – twelve miles through up to three foot of ash.

Each day, at the Observatory above, brought different

experiences; frantic seismic action followed by two days in which the landscape showed 'not one note of colour'. By the 18th, they noticed the effects of the gas – difficulties in breathing and the 'indescribable feeling of oppression'. Local people were becoming desperate, struggling to get back to their houses, crops and livestock; some forty appeared, roped together, hoping to use the barrack building nearby but were forced to take shelter at the Observatory. One old man and one young bronchial sufferer died. By 22 April, all was finally quiet.

No one had ever described a volcanic eruption from such a close viewpoint, effectively within the zone of the volcano itself. Perret's book, with its photographs and graphic descriptions of the two weeks he had spent holed-up in the Observatory – 'thrilling days' – gave it all an instantaneous reality that was entirely in tune with the spirit of the new century, so fond of immediacy and events recorded on newsreels, in which explorers and adventurers broke the limits of the possible. But as Perret stated, Professor Matteucci had been the hero of the event and he was to be the martyr, too. The Director's prolonged stay at the Observatory, during the pre-eruptive phase and throughout the actual eruption, brought about his death soon after, from ash inhalation.

Once calm returned to the mountain a succession of distinguished visitors, scientists and grandees alike, began to make the ascent, appreciative of the professional dedication of the Observatory staff and helpers, and eager to see the transformation and remodelling of the Vesuvian landscape. One of the first on the spot was Dr Johnston-Lavis, who for some years now had been practising medicine in the south of France, but had been irresistibly drawn back by recent events.

Vesuvius, he discovered, had lost some 115 metres off its cone, rendering his wonderful map obsolete.

7

....

VESUVIUS AND THE WIDER WORLD

Thomas Cook, 'the Napoleon of Excursions', led his initial tour to Rome and Naples in 1864. The first edition of Murray's Handbook to Southern Italy had been published eleven years earlier. Many more followed, with revisions. The preface to the sixth edition alerted readers to a couple of dramatic changes – political and physical. The Neapolitan monarchy had been 'blotted off' the map as the editor put it, while Vesuvius itself was 'remarkably altered by the eruption of 1868'.

As the geologists in the new Observatory focused on the volcano and its behaviour, and southern Italy joined unified Italy, a growing public, new to overseas travel, was heading to Naples. Cook was astute and steered his clients to familiar scenes, giving them the reassurance of a knowledgeable guide and the company of the like-minded. Now, like some predatory beast with its trophies around its feet, Vesuvius could also show off its victims. During excavations at Pompeii in 1863, Giuseppe Fiorelli, the ambitious young superintending archaeologist, found a shallow cavity. Filling it with plaster, he discovered that he had made perfect casts of a group of fleeing Romans, including a 'pregnant' (or bulkily dressed?) woman, captured in their final moments as the deadly volcanic ash caught and then incarcerated them. As the succession of poignant 'corpses' went on display in the museum at Pompeii, interest grew. Even to this day, it is these contorted human bodies, often in obvious agony, that visitors, in some millions, flock to see. Pompeii's selling point is, crudely put, mortality as spectacle – two-thousand-year-old mortals caught in their death throes. Since early days Herculaneum, left unexcavated in the nineteenth century, and with no exhibits to match these (although in 1981 a number of complete skeletons were found close to the ancient shoreline) has tended to be less visited despite its greater legibility as a Roman town and its compelling atmosphere.

For those fortunate early Victorian expatriates who had chosen Naples, with its balmy climate, for the good of their health, like the affluent Romans did when they built their sumptuous villas around the bay, the invasion by 'Cook's Tourists' unleashed their rank snobbery. Yet most of these new travellers were discerning enough to benefit from the services of expert guides and the reassurance offered by local agents who spoke the language. Their tickets allowed them to travel by a combination of train, steamboat and, in the early days, the uncomfortable and overcrowded stage coach known as a diligence – an utterly unsuitable vehicle in which to face an Alpine pass. Grand Tourists, Romantic writers and artists had been spurred on by their aesthetic and literary preconceptions but Thomas Cook & Co.'s clients may have been unaware that they were enjoying a Sublime experience as they made the terrifying crossing of the Alps over the St Gotthard Pass. When the travellers finally arrived safely in southern Italy, they found themselves in a landscape of classical antiquity already familiar from a welter of illustrated anthologies and books such as Samuel Rogers' *Italy* (with prints after Turner) rather than from classic literature. Vesuvius was always a favourite destination, boldly lettered onto Victorian school maps, one of a limited number of towns and important sites that could be fitted legibly onto the spindly Italian peninsula.

Potential visitors were well prepared for Vesuvius by colourful travel journals and descriptive pieces such as the account of the 1855 eruption published in Dickens's *Household Words*. Henry Wreford was also the *Times* correspondent in Naples; but for the magazine he wrote a fast-paced and entertaining report more like a modern piece to camera on prime-time television news.

Wreford did not reach the volcano until a week after the eruption (until then it had been judged too dangerous for public access). The cost of carriage hire had quadrupled to meet demand but he and a friend joined an immense wave of traffic in which even carriages 'of high degree' were forced to slow down and go single file, while the colourful local painted carts, *corricoli*, 'shoot by us like a mail-train'. One driver, standing up at the reins and wearing a Phrygian cap (Wreford was, perhaps, hinting at the revolutionary atmosphere in Naples), carried fifteen passengers, including the obligatory fat priest, three of whom were suspended in a net underneath, between the wheels. This awkward freight was, he claimed, pulled at high speed by a single horse.

Pedestrians streamed along the roadside like continuous moving walls of humanity, while stalls had popped up, like snails in the rain, to sell dried peas and beans, melon seeds and black olives – sustaining snacks for those with a long and arduous night ahead. The locals made their purchases and then ran to catch up with their fellows, shouting and laughing. Wreford, not overly enamoured of the Neapolitans, imagined the jokes to be at their expense 'if I read their looks and signs aright'.

At Resina everyone was prey to a bevy of opportunistic guides, porters and torchbearers, bearing down upon their carrion like hungry raptors. The cigar-smoking Englishmen must have looked rich pickings and though Wreford protested that they did not need help, a persistent guide jumped onto the back of their carriage and sunk it axle-deep into the debris. Stung by the continual derision of taunting locals ('the mountain will stop for them of course; don't you see they are English!') they got out, leaving their vehicle to add to the mad traffic jam at the Hermitage.

Above it all, set apart from the torch-lit mayhem it had caused, soared the newly-minted cone; 'like a huge giant, whose side was seamed with wounds, from out of which poured forth his very life-blood'. Wreford was intrigued by the evidently determined course of the lava, two distinct streams coalescing into a single river of fire, cooling as it made an orderly exit between Monte Somma and Vesuvius.

The bulk of the crowd turned left, but the intrepid pair took their own course and began to walk on lava that, Wreford assumed, had only just been thrown out of 'the bowels of the mountain'. The heat and sulphurous smells were excruciating. Their boots and clothing so singed, they discarded their footwear for more rudimentary protection and rolled their trousers up to their knees. Jumping from one lump of scoria to another 'like dainty cats shod with nutshells', they found themselves on the brink of the lava river. For Wreford, the impact was profound and thought-provoking, for he had reached the ultimate goal of every volcano watcher. What, he mused, was the source of such implacable power 'conveyed by the silent, majestic, irresistible course of the miraculous stream'? He imagined that his feelings might be like those of primitive man faced by a steamship or mail train.

The mountain looked like an upside-down colander, fumaroles splitting open on every side; 'no two people behold the mountain under the same aspect, so continual are the changes'. Onlookers were silhouetted against billowing cushions of smoke and painted red, 'like the presiding demons of the scene' – in fact, just as Pietro Fabris had shown them a century earlier. In reality these satanic figures were engaged, noisily, in mundane activities; cooking eggs, lighting cigars or fishing up lumps of lava onto which to impress coins with a special stamping tool – instant home-made souvenirs. Others were picnicking where they could or changing their shirts after the arduous climb. Wreford's report neatly caught the meeting of the mundane with the stupefying. A few people seemed to share his amazement at what they were witnessing; 'Judgement of God!' one muttered.

Despite the hour, the crowd (he estimated it in tens of thousands), was still washing up and down, 'a madness had seized on everyone, and no wonder'. At almost four in the morning they passed King Ferdinand II and his family still making their way up by torchlight. As Wreford and his friend headed down, a cordon of soldiers appeared behind them. He did not make the point, as he might have done, that the increasingly illiberal monarch had now succeeded in making Vesuvius out of bounds to his people.

Wearily reaching the city, Wreford found unshaven café owners dispensing welcome black coffee while the 'sambuca and spirit boys' supplied the demand. Out in the bay, white-sailed fishing boats were bringing in a new haul and life continued uninterrupted, the fresh morning rhythms unaffected by the extraordinary nearby drama. So vivid and immediate, Henry Wreford's journalistic description of those ten or so hours on the exploding mountain must have made disquieting, if thrilling, reading in the stuffy drawing rooms of England in mid-summer.

Once the Mont Cenis Tunnel opened to trains in 1871, the journey to Italy became far less testing and much more mundane. Clutching their copies of Murray, Baedeker or Cook's guidebooks, the new breed of traveller had plenty of practical and sensible advice to hand on what to do and how to do it. Gone was the frustration of reading about new activity at Vesuvius in a London gazette, tantalising for the certainty that it would all be over by the time they arrived. In 1872 some of the early passengers to reach Italy through the new tunnel were lucky enough to get to Naples and find Vesuvius obliging them with a major eruption.

Some visitors arrived by sea, as they still do, taking a few days on shore before they cruised or travelled onwards. These included servicemen on their way to and from duty. My own paternal grandfather, who died long before I was born, was an artillery officer whose transport ship anchored at Naples in 1880 in order to give the troops ground leave. While on shore, the young soldier bought a series of panoramic photographs rather than the mass-produced gouaches that had been on offer until a few years before. He pasted the large black-and-white views of Vesuvius, Naples and Pompeii (the

28. This group portrait is dated 26 April 1885 and shows heavily dressed
tourists, resting before or after their ascent of Vesuvius, which is smoking
away behind them. There is just one woman in the party, and at least five
guides, identifiable by their staves, tanned faces and flat caps.

latter carefully taken out of hours, not a tourist in sight) into the pages of his album. Soon after, the terrible eruption at Krakatoa in 1883 intensified interest in volcanoes.

The old painted panoramas had now been overtaken by the photographic panorama. In the continual search for the new and the gigantic, ambition sometimes overtook practicality. In Berlin in 1903 a multi-segment view of the Bay of Naples was shown, printed on a single sheet of photographic paper five feet wide and thirty-nine feet long. It had to be developed outdoors on a moonless night. The old ways and the new techniques came together neatly with the introduction (in the USA) of the Electric Cyclorama (and the Electrorama) where photographs were projected at immense scale onto the walls of a rotunda, usually one that had been originally built for painted panoramas.

In Naples the growing crowds stoked a flourishing souvenir trade. Coloured postcards, more like eighteenth-century aquatints than twentieth-century Kodachromes (and if possible ornamented with a prized Vesuvian postmark), were posted home to stir the curiosity of those yet to set out. Whenever the volcano had been active, visitors could have a coin of their choice pressed into baking-hot dun-coloured lava, as Wreford had seen it done. Failing that, there were stalls stacked high with pottery embellished with scenes of Vesuvius at full throttle – or little ornamented snuffboxes, pickle dishes or painted fans, all puns on heat and fire. Richer customers preferred gold-mounted 'lava' cameos, sometimes strung into a bracelet which, despite their claim to volcanic origins, had been carved from local limestone. The term 'lava' made them far more desirable. Portable memorabilia had now replaced the expensive pieces of furniture decorated with volcanic scenes copied from Fabris's prints – the items shipped home by the Grand Tourists along with antiquities and portraits of themselves and their friends on the 'classic ground' – as completely as the new breed of visitor had replaced the old. Similarly, glass items made from volcanic ash are sold as popular souvenirs of Mount St Helens, in Washington State, which erupted in 1980.

Every time it shuddered, the popularity of Vesuvius as a destination soared, for the late nineteenth century was another period of gratifyingly regular, but happily not catastrophic, activity. Naples, Capri and the resorts along the Amalfi coast were the fortunate recipients of a bonanza. Probably few of this (still largely European) public had read the classic texts such as Pliny's letters to Tacitus or Strabo's careful description. Not many may have grasped the significance of the numerous ancient, evocative sites scattered across the area. Everybody, however, knew about the catastrophe at Pompeii, the melodramatic embers of the story having been so well fanned by Edward Bulwer Lytton's *Last Days of Pompeii* and continually refuelled by equally histrionic words, performances and images.

Sir Lawrence Alma-Tadema, the prolific painter of hyperrealist scenes from classical antiquity, first visited Pompeii in the early 1860s and built up an unrivalled private collection of archaeological photographs, giving him a rich source of authentic detail. Alma-Tadema also bought a photograph of the 1872 eruption, taken at 3.30 p.m. on 26 April, one of the earliest ever taken. But soon the baton passed from paintings, prints and even photographs to celluloid. The first of many films titled *The Last Days of Pompeii* was made in Italy in 1908. It was a story peculiarly well suited to the purposes of directors of silent movies, with their need for mute, unmistakable, drama. The next version, made in 1913, provided a heightened atmosphere of doom, as a wash of red spread across the screen when the volcano erupted. An American production in 1935 included Pontius Pilate on the cast list, even though Basil Rathbone's character had been long dead. The film included a disclaimer about the accuracy of the story.

The journey from the city to the volcano had become far easier and quicker during the nineteenth century. The king had opened a railway line linking Naples to Portici as early as 1839. A new generation of social realist landscape painters (including Giuseppe De Nittis and the local Scuola di Resina) set out to capture quite a different face of the city, modern and efficient, to set against its perpetually picturesque backdrop. Later came a tram service and in 1880 a funicular to solve the arduous ascent of the cone – a distance of less than a mile. It was a novelty and injected a nicely festive spirit to the volcanic attraction, since funiculars were more familiar humming up and down the cliffs at seaside resorts or ascending to fairytale hilltop castles. Here the idea was a Hungarian engineer's. To begin with there were just two carriages, Etna and Vesuvio, carrying fifteen passengers perched on wooden benches, and a guard for the journey of twelve minutes' duration. The funicular was a double track monorail.

The funicular became part of Thomas Cook & Son's empire when John Mason Cook, who took over the company from his father, bought it from the original promoters in 1887. He decided to face down the notorious guides who found that their closed shop was, for the first time in centuries, under attack. They fought back, severing the line, torching the cabins or throwing them down the mountain. Cook's winning ploy was to close the funicular for six months, allowing him to smoke them out and negotiate a lower fee for their now-limited services. He modernised the workings of the funicular and reopened it in 1889. Southern Italy was flourishing; Francesco Crispi, a Sicilian of Albanian extraction was the first prime minister of modern Italy to hail from the so-called *mezzogiorno* (Southern Italy), holding office in the late 1880s and again in the mid 1890s. He was habitually compared to a volcano.

Thomas Cook & Co. rendered Vesuvius accessible but very much on their own terms. Anyone could come within easy distance of the volcano but if they wished to ascend to the crater, they had to commit themselves to Cook's timetable, travel arrangements (involving a first stage by horse-drawn carriage) and the services of the now-disciplined guides, brought to heel after having 'been accustomed for generations to practise extortion upon travellers' as *Baedeker's Guide* put it.

In 1903 Thomas Cook & Co. dramatically improved access by opening 7.7 miles of new electrified railway line from Pugliano (now Ercolano) to the foot of the funicular, engineered in a variety of ways to deal with the steepening gradient, although travellers remained in their seats throughout, as the gauges and gearing systems changed beneath them. Now that many more passengers could get there, Thomas Cook's empire expanded to include a hotel, the Hermitage, built blithely close to the volcano on the knoll near Eremo, the site of the Observatory. It was advertised as offering wonderful air, both transparent and pure.

Even though the 1906 eruption, more assertive than many, ruptured the funicular and buckled the railway line, the latter

29. When the funicular was opened in June 1880, there were just two
carriages, Etna and Vesuvio, the latter seen here with passengers whose
expressions suggest they might concur with John Mason Cook, who travelled
on the railway despite 'the unpleasant look it had'. Cook saw the potential of
the funicular for tourism and a few years later his family firm, Thomas Cook
& Co rescued the ailing line and carried out many improvements. They even,
bravely, built a hotel on the side of the volcano.

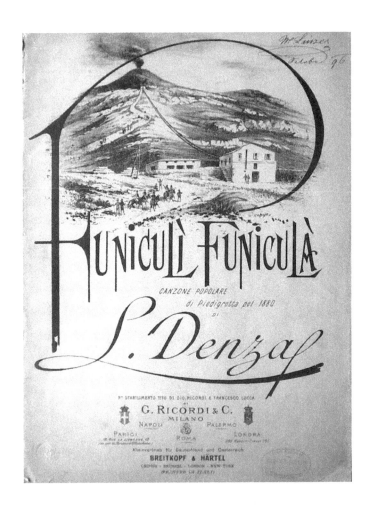

30. The cover of the sheet music for *Funiculì Funicolà* showed the station and the line running to the summit of Vesuvius in all their glory.

remained intact beneath the ash, as did the hotel, which soon regained its vaunted excellent air quality. The funicular took three years to repair and in the meantime, a zigzag track was made so that visitors could go up on horseback. Numbers continued to increase and the company built on its success. In the summer, special late-night excursions were offered since the new enlarged cabins had electric light (and curtains). The modern tourists, like their predecessors, were thrilled by the magic of moonlight on the mountain, the distant lights on the fishing boats and the lighthouse oscillating like fireflies over the Bay of Naples, enhanced by a modest suggestion of menace, as an occasional pregnant wisp of smoke eddied up from the crater.

And there was music too. Compared to the thin gruel offered by early-nineteenth-century opera composers, even responding to the dramatic events of AD 79, 'Funiculì, Funiculà,' the song written by the local journalist Peppino Turco and set to music by Luigi Denza to celebrate the 1880 opening of the funicular, was a runaway success. With its lilting hint of Neapolitan folk tunes, its sentimental lines about ungrateful hearts and the consuming fires above, coun-tered by the reprise 'let it go, let it go, funicolì, funicolà'; it was an immense popular hit from the start. Translated into many languages it sold untold quantities of sheet music as well as thousands of the discs and rolls that fed parlour musical boxes and brought the keys of the pianola jumping to life in front rooms around the world. Every famous Italian singer had a go at it – from Mario Lanza to Luciano Pavarotti.

By the thirties, tales of Vesuvius, and especially of Pompeii, were well-aired favourites on the cinema screen. But the travelogue, slotted in before the main feature, was a new idea, introduced to appeal to an audience of armchair globetrotters. It proved a tempting area for experiment, particularly with the brand new technology, Technicolor. The young cinematographer Jack Cardiff arrived in Naples with his team and their massive, priceless, camera in autumn 1936 with the object of making another film in their 'World Window' series, to be called *Eternal Fire*. The plan, initially a rather tame one, had been to contrast the 'sparkling serenity' of the area contrasted against the distant threat of the volcano but so much smoke was pouring out of the crater that they realised they had arrived at a crucial moment. Now their objective was to get as close to the action as possible.

In his autobiography, Cardiff described how they drove up as far as they could and then, with the essential help of some thirty guides, climbed to within a hundred yards of the crater. There the ground was so hot underfoot that the team had to keep constantly moving while remaining alert to the threat from overhead, since lumps of molten lava were being ejected only to drop straight onto them. Three of the team concentrated on keeping the elephantine camera safe on its tripod and the fourth man hauled the battery along. They stumbled through clouds of smoke and sulphurous vapour, shot just a few seconds of film and then beat a rapid retreat. They repeated this manoeuvre until they had what they needed. At one point the tripod legs melted and it looked as if the fabulously expensive equipment might be stuck fast. (When the Icelandic volcano erupted in spring 2010, a presenter for the TV motoring programme *Top Gear* attempted to drive over the red-hot material, only to find himself welded to the spot by his tyres. The authorities were not pleased.)

In old age Cardiff, one of the most distinguished camera-men in the history of cinema, confessed that he had slowed the film speed down from 24 to 12 frames a second in order to make the lava flow faster and over-exposed it to make 'wine-red lava more orange'. The landscape painters of eight-eenth and nineteenth century Naples were, of course, not the only ones to profit from artistic licence, subtly manipulating the image, when so much was expected of the results. Early photographs were sometimes 'improved'. Even the estimable Dr Johnston-Lavis doctored his images of Vesuvius, scat-tering debris onto the surface of his glass plate negatives to simulate a shower of volcanic bombs hurtling up out of the crater. He wanted to engage a lay audience with the drama of Vesuvius and to catch a moment that he had witnessed but unfortunately failed to photograph in time. Although Johnston-Lavis's own objectives were primarily scientific, he never under-estimated the importance of the volcano's visual impact or its popular appeal.

But even he could never have achieved what Jack Cardiff managed. When the team returned the following day, now able to film at a safer distance, they were able to take advan-tage of some remarkable props. To Cardiff's amazement, the museum authorities at Pompeii had agreed to allow him to take several of Fiorelli's plaster casts up onto the mountain. The crew began to turn the camera above the sad victims of AD 79, dramatically shooting them lying on the fatal lava, to be seen 'through whirling smoke'. Cardiff had earned himself an initial stern reprimand for being unnecessarily foolhardy with the precious new equipment, having burnt the tripod legs and damaged a lens prism in the intense heat, but once the studio had seen the rushes, they cabled him immediately. His little film was the most exciting material they had ever seen.

⧗

For those whose minds were allusive and suggestible, Vesuvius remained everything it had always been. Sigmund Freud's experience of the volcano, active when he saw it in 1902, reminded him of the god of the Jews. With 'a smoke cloud by day and a fire cloud by night' it was, he reminded his son Martin in a letter many years later, 'just like the God of Exodus in the Bible'. But Freud's interest also extended to the sexual metaphors that could be extracted from volcanic activity and, from that, the Surrealists, or at least the men in the group, embraced the imagery with gusto. The 'archetypal' Surrealist film, *L'Age d'Or*, directed by Luis Buñuel and scripted by himself and Salvador Dali, with Max Ernst playing a part, was shown in 1930 before being withdrawn, its anti-authoritarian messages considered deeply offensive (be that church or state, in France or in Spain) and its imagery implicitly disturbing. It was not seen again for some fifty years. In the hour-long film, Buñuel took volcanic imagery to extremes, mixing eroticism and scatological fantasies. In 1939, André Masson's painting 'Gradiva' included an ejaculating Vesuvius shown as the background to a mangled female figure; he took his subject from a short story about an archaeologist sexually aroused by an antique marble statue of a woman, a story which Freud had unpicked many years earlier. Max Ernst returned to the volcano as subject just after the war.

31. A press photograph that effectively captures the turmoil of southern Italy in March 1944, so vividly caught by Norman Lewis in his written account *Naples 44*. Vesuvius erupted soon after the Allies had landed. Here US army jeeps speed along the track and behind them the volcano is spewing out an immense ash cloud. American vehicles helped evacuate people in the areas most vulnerable to the lava flow.

Post-war directors in the Italian neo-Realist cinema such as Roberto Rossellini took up the symbolism, using the broken fabric and ancient remnants of Pompeii and its nemesis Vesuvius to convey strong, even subversive, emotional and political messages.

The outbreak of war ended Surrealist exploration just as it did travelogues and tourism. Those who arrived in Naples now came on military business and it was in September 1943 that Norman Lewis, a young intelligence officer, serving in the Field Security Police, landed. His first wife was Sicilian, so that he was exceptionally open to the unnerving mixture of style and mild anarchy he encountered in Naples even in these months of stark desperation when it must have most closely resembled the desperate city of the *lazzaroni* in which William and Catherine Hamilton had arrived almost two centuries earlier.

On 19 March 1944 Lewis was distracted from his duties by, literally, a bolt from the blue. 'Today Vesuvius erupted. It was the most majestic and terrible sight I have ever seen, or ever expect to see.' He was in Posilippo perhaps, he surmised, on the very same spot that Pliny witnessed the scene, as well as Nelson and Emma Hamilton (like most people, he easily forgot poor Sir William). Ironically enough, the Allied forces had at first thought the noise was from the detonation of an immense bomb.

The menacing and seemingly static pine-tree shaped cloud, which by evening had risen to a height of tens of thousands of feet, seemed solidly three dimensional, 'moulded on the sky'. The scene was unearthly in its scale and stillness. Then by night came a 'terrible vivacity': the lava, trickling down the mountain, was inscribing fiery symbols on the water of the bay while overhead the crater was spitting 'serpents' into the bloodied sky as it continued to pulse with lightning. Like Wreford, he watched the opening acts of the drama from the city.

The next day was still and monochrome, the air and the ground thick with ash falls. Military installations had to be checked, especially at Portici and Torre del Greco, so vulnerable to the mountain, and Lewis was given the job. He drove, in a 'slow, grey, snowfall' to visit an eminent seismic expert, Professor Saraceno, who was – unsurprisingly – 'pleasantly excited at the prospect of the vindication of certain of his theories'. He was convinced that part of the crater wall had earlier collapsed inward, causing a temporary seal and build-up of pressure, suggesting another greater explosion to follow. If so, 'I got the impression that he would not be wholly dismayed,' Lewis recorded. The next day the eruption did, as Saraceno had predicted, intensify and Lewis was sent to check on the village of San Sebastiano, the most severely threatened of all the villages in the vicinity. Getting there was hard work. The tyres of his vehicle skidded in the sticky ash, small *lapilli* (the diminutive of *lapis*, a stone) rained down as well as *bombe* (larger rocks), while directly overhead the deadly cloud was suspended 'full of swellings and protuberances, like some colossal pulsating brain.' It was astonishing that San Sebastiano had been built on this site at all, squeezed between the immense lava fields of the 1872 eruption, on a tip of land that had been spared. As Lewis put it, 'any outsider would have predicted the town's eventual destruction as a matter of mathematical certainty … [but] civic permanency is a matter of religious faith.' The villagers had defied the dun-coloured desert around them and the grey cone of the volcano

32. Villagers make for shelter as Vesuvius erupts in March 1944. People use umbrellas, heavy coats and any protection they can find against the cloying ash and debris that was falling and turning daylight to dusk, just as Pliny had described.

to their rear (the houses looked fixedly west to Naples), by indulging in the brightest paint to hand, filling their window boxes with flowers and even adding gaiety to the street scene with political posters and banners in strident primary colours.

Lewis arrived to find 'the lava was pushing its way very quietly down the main street, and about fifty yards from the edge of this great, slowly-shifting slagheap, a crowd of several hundred people, mostly in black, knelt in prayer.' Sacred images, reliquaries and banners, largely those relating to the saint after whom the village was named, were their sole protection from the huge sluggish tide of material. All they did, apart from the occasional hysterical individual running forward with a religious token to remonstrate with the lava wall, was to continually scatter holy water and swing censers. As this 'slow deliberate suffocation of the town under millions of tons of clinkers' continued, at the rate of a few yards an hour, a surreal sight appeared in the shape of an entire church cupola, neatly severed from the walls of the building below, being borne along on the top of the thirty-foot-deep river from Hades. When the lava met a building, rather than engaging in a King Kong-like confrontation, it appeared to be ingested whole, attended by faint grinding noises.

So desperate were the villagers they had almost broken a religious taboo. Their own modest patron saint had failed to protect them and so, down a side street, veiled by a sheet lest it give offence, was laid an effigy of San Gennaro who, if all was to fail, might be their insurance policy. But as he returned, at that moment, to the main street, Lewis saw that the lava had slowed and was now at a standstill. Half the town had been saved and their saint's honour was intact.

Now the saints had to prove themselves. So suggestible

33. San Gennaro awaits his moment in March 1944 attended by a stoic village
woman and a couple of lads more interested in the camera than the effigy.

were people that if San Gennaro did not perform next time, it might be interpreted as the fault of the Allied forces. 'The war has pushed the Neapolitans back into the Middle Ages,' Lewis wrote. 'Churches are suddenly full of images that talk, bleed, sweat, nod their heads and exude healthy-giving liquors to be mopped up by handkerchiefs, or even collected in bottles, and anxious, ecstatic crowds gather waiting for these marvels to happen.' The newspapers fed the fever; San Gennaro even had an understudy. If his blood failed to liquefy, a phial of the blood of St John was kept in San Giovanni a Carbonara which bubbled up every time the gospel was read. Lewis put the hysterical atmosphere down to nervous exhaustion, and now 'mass hallucination has become a commonplace, and belief of any kind can be more real than reality.'

Lewis's account of the actual eruption, a masterly description with his almost filmic emphasis on the sounds and pace of the terrifying lava incursions, can be compared to actual film shot by American service personnel (including fighter pilots). For the first time ever, this eruption of Vesuvius was shown as it happened, the long series of unfolding events now edited down to several minutes of thrilling black-and-white footage on YouTube. Much of it is like a grisly, all too immediate, revisiting of seventeenth- and eighteenth-century images; silhouetted figures (and sometimes close-ups of sad, weary faces) watching from a distance as their vines turn to charcoal, or hapless villagers parading effigies of the Virgin and the saints around their streets. The moving images are compelling in their immediacy; the towering cloud of ash rolls on, over and over, up and up, while the relentless river of lava (like jam nearing its setting point) creeps on down, crushing buildings like walnuts. One crack in the masonry and the entire structure implodes. Even the equipment of war took its share of the devastation, with heavy losses of American planes on the airfield at Pompeii. Almost eighty were destroyed – burned, melted, cracked or otherwise disabled – by the falls of hot ash and tephra. Vesuvius had become the Axis's newest secret weapon. Hardly eighteen months later the Americans dropped the atom bomb on Nagasaki; the mushroom cloud there was 18 kilometres high.

This eruption, an episode already vanishing into the mists of history, is now the most recent volcanic event at Vesuvius. Reading about those days in early 1944 is a prompt of a terrifying kind. When will Vesuvius come to life again, as it must? How will Italy handle the next episode? Can it cope with even a 'routine' effusive eruption let alone the devastating, and not unlikely, possibility of a deadly pyroclastic flow, a fast-moving tidal wave of burning gas and fragmentary material from which there will be no shelter and of which there will be no warning? The lack of respect for the process of the law and regulatory frameworks, in particular regional and local planning which have been made toothless in the face of rampant *abusivismo* (the word coined specifically for irregularities in the planning system), is endemic. Urbanisation of the entire district means that the lower slopes of Vesuvius are now matted by a dense web of illegal buildings – commercial, retail, residential and even public amenities. There is a major oil refinery and a brand new regional hospital (albeit built to the highest seismic standards) abutting the Red Zone, the area considered to be most at risk from another eruption.

Two municipalities, Torre del Greco and San Giuseppe Vesuviano, lying to the south and south-east, the areas most likely to be hit by immense lava streams or unexpected 'flank' eruptions breaking out through new apertures, continue to actively expand into volcanic hazard zones. After the devastating eruption of 1794 the elderly Sir William Hamilton, who was by then as seasoned a volcano-watcher as anybody could be, was shocked by the disregard for the safety of the local population. He wrote to Sir Joseph Banks; 'What business have people to build at the foot of a Volcano & yet they are going directly to build at Torre del Greco? It contained 18000 Inhabitants and the remaining houses can not contain a quarter of that Number.' Just fifteen months later he reported that the town was almost entirely 'rebuilt on the lava altho it burns their fingers when the[y] dig into the Scoria to lay the foundations.'

Nowadays such infringements are usually blamed on the Camorra, for it is always easy (and often justified) to shift the blame to the perpetrators of organised crime, but bureaucratic stalemate, inertia and petty corruption are culpable too. In the entire area of the Circumvesuviano (as the circular suburban railway line that runs like a loose cummerbund around Naples is called), three danger zones have been designated, in ascending order; in C, the quantity of ash would endanger respiration (and might bring down roofs); in B, the danger would be that of injury from severe burns; but in A, in the case of a pyroclastic surge there would be no chance of survival whatsoever.

In reality, the gulf between what John Dickie and John Foot in their introduction to *Disastro!: Disasters in Italy since 1860* term 'real Italy' and 'legal Italy' seems an unbridgeable chasm. The prevailing, understandable attitude is fatalism, similar to that around the San Andreas Fault in California, but unlike there, the Neapolitan infrastructure is utterly inadequate for a mass evacuation of several million people within a few days. Improvements to the road that embraces the lower slopes of Vesuvius, the SS 268, convince no one that it can carry traffic on the scale required if the citizens of the area were ordered to leave en masse.

At the macro level Italy has a National Disaster Plan covering Etna, Vesuvius and the Po Valley. As David Alexander writes, Italy (the European country judged most at risk of natural disaster) has developed sophisticated disaster preparedness, with a 'cascading system' of emergency command. It is only in Italy that the profession of 'disaster manager' exists. Emergency planning, on paper at least, is well developed. Any eruption of Vesuvius will put at least 650,000 and at most 3.1 million people at risk. The volcano is in the sights of those given the task of dealing with an eruption yet, as Alexander points out, the actual response has descended into 'a dog's dinner of conflicting strategies, policies and legislation'.

In late 2009, the principal scientific agencies (like the disaster emergency agency itself) were faced by the threat of privatisation. The staff spent their Christmas holidays sitting out on the roof of their offices in Rome. The fear is that cronyism, if not corruption, will undermine the ability of these bodies to react in the national interest while political spin muddies the waters further. When the deadly mudslip of spring 1998 hit Sarno and the surrounding villages in a heavily deforested, hilly area south of Naples, it was convenient, particularly at a time when the far right (the nationalist Lega Nord) was flexing its muscles, to ascribe the catastrophe to the fecklessness of southern Italian planning legislation and regulation.

The earthquake at L'Aquila of 2008, and its aftermath of humanitarian disaster exacerbated by mislaid funds, is more difficult to explain away on those grounds.

In the areas around Vesuvius, reputable banks will not offer people mortgages or loans for house building, there are no construction licences to be had and no insurance is available, yet the desirability of the position, midway between mountain and ocean, green and salubrious above the urban density, is undeniable. There is a great deal of recent building to be seen on the lower slopes of Vesuvius, both family houses and large blocks of flats. It is worth paying cash (and local taxes) to achieve a toehold on this kind of site and who knows, so the attitude goes, if another entire century might pass without the volcano emitting more than its current tiny vapour trails. As one Neapolitan, born decades after the last eruption and extolling the advantages of this location told me, Vesuvius is, put simply, 'part of us'. This is so much the case that *Il Cavaliere* (Silvio Berlusconi) built an artificial volcano at his Sardinian villa for Ferragosto (15 August) in 2006. It was his 'crazy idea' for thirty carefully selected guests. He was out of office at the time, but preparing to reenter the lists as Prime Minister. The *Corriere della Sera* newspaper quoted an anonymous witness who described the surreal vision, a cone surmounted by a confection of lights, optical fibres and smoke, which seen by night appeared to be erupting, with lava pouring down. Artificial earth tremors added to the effect. It all brings Las Vegas to mind. Fortunately the fire brigade was on stand-by that night.

The slopes of Vesuvius offer a desirable alternative to Neapolitan reality, the greenest of suburbs, its gardens full of laden citrus trees and vegetables still growing well into the winter, thanks to the extreme mildness of the sheltered climate. Any retrospective enforcement action would be the responsibility of central government. The case rests. Recently the regional government tried offering households 30,000 euros as an inducement to settle outside the endangered Red Zone, accepting that residents could return to work in the area. By registering a new address (usually that of a family member) but remaining on the spot, families still claimed the money and took no further action. Back pockets were filled. Even the head of the national civil protection agency is on record saying that the initiative has been a total failure.

No one now lives in the district around Vesuvius in order to gain a vicarious thrill, as the pleasure-seekers of the Golden Mile did in the eighteenth century. Many of the estimated half a million people living in the eighteen communities within the Red Zone, those around the craters of Vesuvius and Monte Somma, are without regular work and are poorly educated. The risk is no longer a real one in the minds of the majority of the locals, for only the very elderly can remember the most recent eruption, that of 1944. Yet a recent study of 400 young people in the immediate area suggested that they were all too aware of what the volcano could do, their risk perceptions (and fear) were acute, but they had little informa-tion and had received no 'hazard education'.

The natural reaction is, psychologists say, the application of 'bounded rationality', a term coined by Herbert Simon to describe an understandable human impulse to seek the best possible solution, rather than the ideal one. Academic studies of attitudes to danger in those living close to Etna have revealed a (non academic) response, something along the lines of *Che sarà, sarà.*

Yet, a measure of psychological adjustment to the realities of an eruption at Vesuvius may have been signalled by the apparently unexpected events in Iceland, and the repercussions of Eyjafjallajökull's resulting ash cloud which ran around the world in spring 2010. Twenty-four-hour news networks, riveting photographic images of volcanic action and a continual stream of factual information from sources like the Volcanism blog, has made many people all too aware. Even the seismographic reports from Vesuvius can now be found online and live. In Iceland, an island with a total population rather less than that of Naples, 70,000 people went to gaze at the spectacle in the interior. An eruption of Vesuvius would lead to a massive flight out of the area, a tsunami of people, while an answering tidal wave would come in, determined to see the eruption with their own eyes. After the eruption in Iceland, Guido Bertolaso, head of the Italian civil protection agency, said that Vesuvius constituted the most urgent of their problems (while also signalling that long-dormant Ischia might pose more immediate volcanic worries).

As the years tick by, the moribund crater with its feeble little emissions, seems increasingly benign. Vegetation is returning to the lava strewn areas around the cone, giving the ground ethereal beauty with vivid acid green moss and misty grey-green lichen, all too much like puffs of smoke. More plans are hatched to replace the funicular, while the replacement one constructed in the late 1980s lies forgotten in the undergrowth, not far from the site of the original. The shabby car and bus park and (charmingly) old-fashioned shop, surrounded by Portakabin lavatories disguised ineffectually behind laminated photographs of Vesuvius in different lights and seasons, is untouched by any of the modern mantras of interpretation or access although an exemplary modern exhibition is housed in the Observatory below.

Meanwhile artists, as they have done for centuries, continue to toy with the effects and engage in the resonances of the place – their impressions now more likely to be gathered online than at first hand. A number of contemporary (as well as art historical) exhibitions have focused on volcanic images or even placed Vesuvius itself centre stage, a galvanising theme. The Italian painter, sculptor and print-maker Mimmo Paladino was born nearby and he is just one internationally recognised figure – with Warhol – to use the imagery of the mountain in his work. In whatever medium an artist works, there is a compelling power in the clash of implacable and intransigent elements against the mutable and the transitory. James P. Graham has been making a 360-degree video work on Stromboli for some time and the counterpoint between the violence of the volcano and the calm of the sea, both on the ear and eye, is highly effective. As the lava meets the ocean, fragments sizzle and steam furiously on impact with the water, like fritters landing in a pan of boiling oil. Heat and cold, fire and ice are compellingly symbiotic. As Dickens wrote, Vesuvius still 'burns away in my thoughts beside the roaring Waters of Niagara, and not a splash of the water extinguishes a spark of the fire; but there they go on, tumbling and flaming night and day, each in its fullest glory.'

Recent research has uncovered the existence of a 400-square-kilometre magma reservoir beneath Vesuvius stretching far out into the sea, lying at a depth of some eight kilometres. This discovery has encouraged the use of seismic tomography, a process analogous to the CAT scan used in medicine. Manmade explosions are set off, at sea as well as at considerable distance on land, in order to generate seismic waves which can then be traced, read and compared as they travel between the seismic stations, providing a multi-dimensional picture of the current condition and behaviour of the magma layer. Devices called tiltmeters are used to see whether the magma is flexing or inflating – there is continuous measurement of the gases to see if anything is occurring. In Naples alone, one hundred and twenty people are employed in surveying the seismic behaviour of the earth's crust in this location.

Fewer than that number came to see the miraculous liquefaction of the blood of San Gennaro taking place, as usual, in the saint's chapel at the Duomo in Naples on 16 December 2009, the anniversary of the 1631 eruption. Perhaps that was as well; after a relay of priests had rocked the phial of dark sludge to and fro, morning until evening, each as uninterestedly as if he were listening to a particularly pedestrian confession, there is no blood or movement of liquid to be seen, even by the end of the day. The anticlimactic ceremony ends with a concert of newly written cello music (hard on the ear) and some devotional readings. By now, even fewer people are present and the 'zei' (the aunts), just a handful these days and in this season, have gone home to their pasta. We go out for a *pizza vesuviana*. It emerges, bubbling up impressively, from deep inside a glowing wood-fired oven.

In May 1944 Norman Lewis recorded how the ceremony, held in the recent aftermath of the eruption, brought almost all activity in the city and harbour to a halt. So febrile was the atmosphere in the Cathedral that the seating of British and American servicemen (considered likely to be heretics and so antipathetic to the success of the miracle) so close to the altar had angered the congregation. It was a long process that spring, and the arrival of the saint's relatives did little to speed it up. Finally, around eight in the evening, the dry material turned to liquid. The verdict was that it had been a 'poorish liquefaction but better than none at all'. Had it failed, Lewis guessed, there would have been civil disorder on a major scale. Even one of Lewis's most sophisticated Neapolitan friends who descried the medieval nonsense confessed that he was, despite everything, still swayed by 'mass suggestion'.

Air travellers flying in, the plane almost brushing the top of the volcano as it comes in to land, may enjoy Vesuvius glowing in an evening sun, wrapped in a modest slip of cloud, or, best of all, capped by a crisp corona of snow, before encountering the realities of the airport. They may follow San Gennaro's relics threading the ancient streets around the Cathedral. Should the mysterious substance in the phial turn to liquid everyone will be happy, judging Naples spared for a little longer.

The very docility of Vesuvius, the lack of threat or menace except in fading memory, caricatured on menus with their ersatz *pizze vesuviane* or by repetitive images on souvenirs, rising out of snowstorms or on pottery ashtrays, has come to lessen its allure. In the volcano's current quiescence, the surroundings buried by a shabby tidemark of derelict bars and cafés, fly-tips and broken funicular track (this is now a National Park, but not that you would notice) it can do no more than offer visitors a spectacular view over the Bay of Naples from the highest point and then a glimpse of a

34. Vesuvius as pleasing graffiti, painted on a wall in Naples.

wisp or two of steam rising in the crater, hardly more than the exhaust from an old Vespa. Yet this passive demeanour is cruelly deceptive. Dotted about the surface of the mountain, as well as further afield and out to sea, an astonishing array of devices is in place, waiting to detect any change in local seismic behaviour. As Susan Sontag wrote in the prologue to her sparkling novel based on the life of Sir William Hamilton, *The Volcano Lover* 'what happens once can happen again. You'll see. Just wait.'

VISITING VESUVIUS
AT HOME AND ABROAD

Vesuvius is reached from Naples on the Circumvesuviano, a surface railway line which, as the name suggests, runs all the way round the base of the volcano. The stop is Resina and in season, coaches take visitors up to Vesuvius. Out of season you may have to spend money on a cab. After the grisly car park, visitors buy a ticket and clamber up the clinker path, and on reaching the crater's edge are accorded the services of a guide (or not, depending on the weather). Down below, about halfway up, is the Observatory which is now an excellent museum – in many ways the most worthwhile part of a visit to Vesuvius under current somnolent conditions. Their website http://www.ov.ingv.it will give you information; in theory members of the public can visit only on Saturdays and Sundays between 10.00 and 14.00, unless in pre-booked groups (or school parties) in which case the hours are 9.00 until 14.00, Monday to Friday. Out of season they seem to take a rather more flexible approach.

Resina is also the station for Herculaneum, less crowded (and in my view, more atmospheric) than Pompeii and over which Vesuvius looms alarmingly. At Pompeii, further along the railway line, is a museum dedicated to Vesuvius, the

35. Visitors are going up Vesuvius in horse-drawn vehicles, to be deposited at the station above, from which they will embark to the summit on Thomas Cook & Co's funicular. Afterwards they will return to the coast or the city for Cook's hotel is not yet built.

Museo Vesuviano 'Giovan Battista Alfano', a scientifically expert monsignor who established his collection in 1911. The current location is in Via Colle San Bartolomeo and it is open on weekday mornings (not holidays) from 9.00 until 13.00. Back in Naples on no account miss the chapel of San Gennaro in the Duomo, as well as the Treasury (ticket). If time and opening hours permit, try to visit one or two of the few surviving villas on the Golden Mile. http://www.villevesuviane.net

OTHER VISITS

The gallery at Compton Verney, an elegantly converted country house in a Capability Brown landscape, can be reached via junction 12 off the M40, or from Leamington Spa, Stratford-upon-Avon or Banbury stations, and specialises in Neapolitan art including a number of fine paintings of Vesuvius. http://www.comptonverney.org.uk

The garden realm of Dessau-Wörlitz is a UNESCO site. It can be reached by train or car from Berlin, a journey which takes around an hour and a half. There are plenty of places to stay in Wörlitz itself but book ahead. It takes two days, at least, to see the full landscape and (at least) the Gothic House, Country House and Luisium, as well as the Stein – the island on which the volcano takes centre stage together with the Villa Hamilton. See http://www.woerlitz-information.de

In London, see the Johnston-Lavis collection at University College. Most of the items are in the Rock Room but don't miss the great map. You must make an appointment. http://www.es.ucl.ac.uk/department/collections/JLC_files.

Finally, at the British Museum, the wonderful Enlightenment galleries which have been installed in the old King's

Library and for which there is a splendid catalogue, place rocks and minerals, geological phenomena and classical antiquity within a context that William Hamilton would have recognised.

FURTHER READING

CHAPTER I

My journey, on paper, to the heart of Vesuvius has been guided by Haraldur Sigurdsson *Melting the Earth; the History of Ideas on Volcanic Eruptions* (Oxford, 1999). An Icelandic vulcanologist he also emphasises and puts into sharp focus the shifting human response to that incredible geological phenomenon, starting with the ancients. He also has a good eye for intriguing images. In a brief section devoted to 'early concepts of volcanism' David A. Young *Mind over Magma: the Story of Igneous Petrology* (Princeton and Oxford, 2009) also conveys man's early attempts to grapple with evidence of volcanic activity with great clarity. I have felt my way towards the landscape of antiquity thanks to several sources; John D'Arms *Romans on the Bay of Naples: a Social and Cultural Study of the Villas and their Owners from 150 BC to AD 400* (Cambridge, Mass., 1970) provides valuable insights into the life of the ancients around their pleasure villas in the shadow of the volcano, closely based upon contemporary written evidence. Tony Harrison *The Grilling* which intriguingly weaves together a number of literary responses, was published in the *London Review of Books* June 2002. Peter Stothard *The Spartacus Road* (London, 2010) ingeniously intertwines the landscape of Campania, ancient and modern, and peoples them to excellent effect. Vitruvius *On Architecture* (translated Richard Schofield, introduced by Robert Tavernor, London, 2009) is the first architectural treatise and accords volcanic material a modest amount of attention. For Pliny and his letters to Tacitus, I turned (again) to Sigurdsson. Robert Harris *Pompeii* (London, 2003) paints a compellingly vivid and atmospheric picture of the days before the eruption, fiction based upon intense research. No reader will forget his evocation of those menacing days and hours as the water ceased to flow. But for a brilliant picture of what was going on there, in the shadow of Vesuvius, Mary Beard *Pompeii: the Life of a Roman Town* (London, 2008) catches every facet of a real world, as it existed before it was 'interrupted'.

CHAPTER 2

I have drawn on the entry for San Gennaro in the online edition of the *Catholic Encyclopaedia*. Edward Chaney *The Evolution of the Grand Tour* (London, 1998) is particularly good on the pioneer travellers and, for my purposes, especially Thomas Hoby. The events of 1631 are retold in great detail in chapter 7 of Alwyn Scarth *Vesuvius: a Biography* (Princeton, 2009) while Sigurdsson keeps an observant, thoughtful eye on the same cataclysm, the most recent pyroclastic event at Vesuvius. The extraordinary proliferation of reports, as well as prints, horoscopes and other ephemera were the subject of an excellent paper, presented by Dr Lorenza Gianfrancesco at the 'Naples crucible of the world' conference held at the Italian Cultural Institute in London in October 2010. For John Evelyn I have turned to my own biography, *John Evelyn:*

Living for Ingenuity (New Haven and London, 2006). The peerless annotated edition of his diary, E. S. de Beer *The Diary of John Evelyn* (Oxford 2000), is in six volumes, although here I drew on Volume 2, *Kalendarium 1620–1649*, which contains the account of his Grand Tour. Sean Cocco 'Natural Marvels and Ancient Ruins: Volcanism and the Recovery of Antiquity in Early Modern Naples', essay in *Antiquity Recovered: the Legacy of Pompeii and Herculaneum* (Los Angeles, 2007) sets the early scene and points towards my (and their) later chapters. For George Berkeley's observations, see 'Extract of a Letter of Mr. Edw [sic] Berkeley from Naples, Giving Several Curious Observations and Remarks on the Eruptions of Fire and Smoak from Mount Vesuvio. Communicated by Dr. John Arbuthnot, M. D. and R. S. S.' *Philosophical Transactions of the Royal Society* 1717, Volume 30, pp. 708–13. For an excellent account of Naples over the centuries, particularly helpful as the city's history becomes ever more complex, Jordan Lancaster *In the Shadow of Vesuvius: a Cultural History of Naples* (London 2005).

CHAPTER 3

The splendid exhibition catalogue edited by Ian Jenkins and Kim Sloan *Vases & Volcanoes: Sir William Hamilton and his Collection* (London, 1996) has been my main authority for this chapter. In addition, Hamilton's standing as a European figure of consequence is emphasised in David Constantine *Fields of Fire: a life of Sir William Hamilton* (London, 2001) while in fiction, Susan Sontag *The Volcano Lover* (New York, 1992), if speculative is remarkably persuasive and offers many insights on Hamilton and his passions. Two academic studies are Noam Andrews 'Volcanic Rhythms: Sir William Hamilton's Love Affair with Vesuvius' in *AA Files 60* (London, 2010) and Bent Sorensen 'Sir William Hamilton's Vesuvian apparatus' in *Apollo*, May, 2004. Jenny Uglow *The Lunar Men* (London, 2002) extends the interests and influence of this remarkable network of Enlightenment men well beyond the confines of their moonlit meetings in the English Midlands. Volumes 24 and 35 of *The Yale Edition of Horace Walpole's Correspondence*, ed. W. S. Lewis (New Haven, 1937–83) include letters to and from Hamilton, while in *Scientific correspondence of Sir Joseph Banks, 1765–1820* ed. Neil Chambers (London, 2006) the relevant letters are found in volumes 1, 3 and 4. Martin Rudwick *Bursting the Limits of Time* (Chicago, 2005) sets Hamilton within a far wider context, that of the immense panorama of geohistory.

CHAPTER 4

Johann Wolfgang Goethe (ed. W. H. Auden and Elizabeth Mayer) *Italian Journey* (London, 1962) was an invaluable source in both the previous chapter and this one. Katherine Wilmot *The Grand Tours of Katherine Wilmot*, ed. Elizabeth Mavor (London, 1992) is a spirited account while I was thrilled to find Martha Coffin Derby's matching account online. (As often in these things, the unnamed source appears to be now offline.) For the Romantic response, I turned to the account in Richard Holmes *Shelley: the Pursuit* (London, 1974) and to Thomas Love Peacock *Peacock's Memoirs of Shelley: with Shelley's Letters to Peacock*, ed. H. F. B. Brett-Smith (London, 1909). Also, for the meeting between poetry and science, Richard Holmes *Coleridge: Darker Reflections* (London, 1998) and *The*

Selected Correspondence of Michael Faraday eds. Pearce Williams, Fitzgerald and Stallybrass (Cambridge, 1971) combine with papers and lecture notes in the archives of the Royal Institution to amplify the ideas of both Davy and Faraday on volcanology. For the shifting fashions in the portrayal of Vesuvius and its visitors see chapter 5, Malcolm Andrews *Landscape and Western Art* (Oxford, 1999). Nelson Moe *The View from Vesuvius* (Berkeley and Los Angeles, 2002) discusses shifting responses and imagery touching southern Italy during the lead up to the Risorgimento and particularly the impact of Leopardi.

CHAPTER 5

Among the many publications on Dessau-Wörlitz very few are in English. *Der Vulcan im Wörlitzer Park* (Berlin, 2005) is so splendidly illustrated that it is worth buying even without the language! A more general account is *Infinitely Beautiful: the Dessau-Wörlitz Garden Realm* ed. Kulturstiftung Dessau Wörlitz (Berlin, 2005) while more detail is given in the 1997 exhibition catalogue *For the Friends of Nature and Art; the Garden Kingdom of Prince Franz von Anhalt-Dessau in the Age of Enlightenment.* The world of homemade entertainments large and small is wonderfully described and detailed in Richard D. Altick *The Shows of London* (Cambridge, Mass. and London, 1978). In addition is Ralph Hyde *Panoramia* (exhibition catalogue, London, 1988) a wonderful window on one particular way of looking at life, near and far. Isabel Armstrong *Victorian Glassworlds* (Oxford, 2008) is a thoughtful exploration of imagery and artificiality. Marguerite Blessington, Countess of Blessington *The Idler in Italy* (London 1839–40) pointed on, in many ways, directly to Charles Dickens' account of his first visit to southern Italy. Jason Roberts *A Sense of the World: How a Blind Man became History's Greatest Traveller* (London, 2006) is the story of James Holman. Nick Yablon '"A Picture Painted in Fire"; Pain's Reenactments of *The Last Days of Pompeii*, 1879–1914' is in *Antiquity Recovered* (op cit) and offers an account of the international art, and taste for, of pyrodrama. Hilary Spurling *Matisse: the Life* (London, 2009) is my source for the evocative scene in Bohain-en-Vermandois.

CHAPTER 6

The useful ODNB entries for George Poulett Scrope and Charles Lyell are by Martin Rudwick. His magisterial *Worlds before Adam* (Chicago, 2008) is the successor volume to *Bursting the Limits of Time* (op cit). Telling the story of a thoroughly engaging man who switched his concerns from this earth to the moon later in life, is *James Nasmyth Engineer: an Autobiography* ed. Samuel Smiles (London, 1883). The website http://www.ov.ingv.it/inglese/storia/storia.htm provides key dates and personalities in the early history of the Osservatorio Vesuviano. Eugene Dwyer 'Science or Morbid Curiosity? The Casts of Giuseppe Fiorelli and the Last Days of Romantic Pompeii' is in *Antiquity Recovered* (op cit). *Personal Recollections from early life to old age of Mary Somerville* ed. Martha Somerville (London, 1874) is a dutiful daughter's account of a remarkable woman. The exhibition catalogue, 'Giuseppe De Nittis: La modernité élégante' (Paris, 2010) gives due weight to the work of this painter who studied and recorded Vesuvius with such care and flair. Another catalogue *Violent Earth: the Unique Legacy of Dr Johnston-Lavis* (London, 2005) celebrates the extraordinary legacy of the British doctor who became a self-taught vulcanologist of great standing. Frank A. Perret *The Vesuvius Eruption of 1906: Study of a Volcanic Cycle* (Washington, 1924) is another account from a similarly self-taught expert. See the excellent website devoted to him http://www.vesuvius.tomgidwitz.com/html/the_hero_of_vesuvius.html.

CHAPTER 7

A fascinating snapshot of the struggles to transport tourists up Vesuvius has been provided by Thomas Cook's company archivist Paul Smith in his article 'Thomas Cook & Son's Vesuvius Railway', *Japan Railway & Transport Review* March 1998. Jack Cardiff *Magic Hour: A Life in Movies* (London, 1996) offers delightful reminiscences at the start of his distinguished and lengthy career. Buñuel's *L'Age d'Or* is now online. Norman Lewis *Naples '44: an Intelligence Officer in the Italian Labyrinth* (London, 1978) is a consummate account of a volcanic eruption and the aftermath. The event can be seen as YouTube footage in various versions. John Dickie, John Foot and Frank Martin Snowden *Disastro!: disaster in Italy since 1860: culture, politics, society* (London, 2002) is a guide to the shoals and rapids of recent Italian history. The publication accompanying James Hamilton's exhibition, *Volcano: the Volcano in Western Art, a Brief Introduction* (Compton Verney, 2010) gives a taste of the excellent exhibition and is, fittingly, dedicated to Maurice and Katya Krafft to whose enthusiasm, films and writing anyone with the slightest interest in volcanoes is indebted. David Alexander *Confronting Catastrophe: New Perspectives on Natural Disasters* (Oxford, 1993) is, as the title suggests, a study of human response in the face of overwhelming natural events. And finally, conveying an innate, effortless understanding of the Bay of Naples, ancient and modern, the beauty and humanity, the ugliness and the beastliness, with admirable brevity and elegance, turn to Shirley Hazzard and Francis Steegmuller *The Ancient Shore: Dispatches from Naples* (Chicago, 2008). They, and Norman Lewis, are unbeatable.

LIST OF ILLUSTRATIONS

ACKNOWLEDGEMENTS

On a particularly dank October day, Uwe Quilitzsch showed me (a non-German speaking stranger) around the gardens and houses that make up Dessau-Wörlitz, even taking me up the volcano and into the inner chamber. He made me promise to return in August for the eruption – and so we did. Thanks to Michael and to Tony McIntyre for joining in a mildly madcap summer holiday expedition. Afterwards, I swapped notes with Tess Canfield (while envying her experience of the eruption on a perfect moonlit night, twenty-four hours earlier) since she knows the site and its history better than I. My previous field trip to Vesuvius was in the footsteps of John Soane with Michael and Susannah. This time I wanted to be in Naples for 16 December, the third of those three important days in San Gennaro's calendar. My old friend Silvia Corsini valiantly sat in the Duomo with me awaiting the miraculous liquefaction, to no avail, and then toiled up Vesuvius in driving rain. But there were pleasures too, including a tour of the Observatory with Massimo Orazi and the enjoyment of out of season Naples, all ready for Christmas.

At the Johnston-Lavis Collection at University College, London, the heart of Vesuvian studies in this country, Wendy Kirk and Emma Passmore were generous with their time and help. Jane Harrison and Professor Frank James helped me in the archives of the Royal Institution, Susan Palmer and her colleagues brought out some treasures of the library at Sir John Soane's Museum, and thanks are also due to the staff of the London Library, the Royal Society and the British Library.

Along the way people have thrown me invaluable titbits. Robert Thorne pointed me to James Nasymyth, and Andrew Saint introduced me to Giuseppe De Nittis. Miles Thistlethwaite gave me a fine print of Emma Hamilton twirling under Vesuvius. Michael Law responded to my *Guardian* preview of 'Volcano' at Compton Verney, suggesting interesting parallels between Burke and Rilke, while Jonathan Keates shared Neapolitan insights and enthusiasms with me over Italian cakes in Soho. Thomas Jones posted a couple of my volcanic blogs on the LRB website. I discussed Charles Lyell with Brenda Maddox and was lucky enough to meet, entirely fortuitously, Professor Martin Rudwick. Professor Demetrios Michaelides tipped me off about early operatic performances involving Vesuvius. In Bexhill, in the editorial office of John Bennett's eponymous magazine, I found myself at the very heart of the world of fireworks, an introduction which came via Philippa Lewis, herself always a mine of useful suggestions. I talked to John Dickie about some of the darker aspects of modern Italy.

A trio of sharp-eyed readers read early drafts. Emma Passmore (then at UCL, now at the British Museum) was my geological guide, Michael Horowitz my in-house style guide while Carrie Boyd Harte willingly took on the role of the ideal reader, and excelled in the role. To all three especial thanks. They cannot be held responsible for any later errors or slippages.

My agent Caroline Dawnay and her assistant Olivia Hunt have been supportive and encouraging. Finally, at Profile I could not have asked for a better, wiser or more experienced pair of editors than Peter Carson and Mary Beard. Thanks too to Penny Daniel and to Andrew Franklin. The latter listened to my (impractical) idea for boxed sets of Wonders of the World volumes with good humour and patience. Profile Books make their writers feel welcome.

February 2011

INDEX

[240]

WONDERS OF THE WORLD

This is a small series of books that will focus on some of the world's most famous sites or monuments. Their names will be familiar to almost everyone: they have achieved iconic stature and are loaded with a fair amount of mythological baggage. These monuments have been the subject of many books over the centuries, but our aim, through the skill and stature of the writers, is to get something much more enlightening, stimulating, even controversial, than straightforward histories or guides. The series is under the general editorship of Mary Beard. Other titles in the series are: